ACKNOWLEDGEMENT

I thank Dr. Harvey Levy for reading an early version of my manuscript and for his useful suggestions.

I also thank Dr. Flemming Güttler for his invaluable tour of the Kennedy Institute in Denmark and his description of how PKU children and their families are treated in his country. I also want to thank Professor Imran Özalp at Hacettepe University in Ankara, Turkey, for providing me with useful information about PKU screening and treatment in Turkey. Professor N. Hashem at the Ain-Shams University Medical Center in Cairo, Egypt, provided me with similar details about how PKU patients are handled in Egypt.

Ms. Elizabeth Emerick of the Maryland Department of Health and Mental Hygiene was extremely helpful in putting me in touch with families in Maryland with PKU children and for telling me about the existence of the Maryland Alliance of PKU Families, Inc. From several meetings with different members of the Alliance, I gained important insights into the parent's experiences in dealing with their PKU children. Meeting their successfully treated kids gave me a very valuable glimpse into the powerful impact that the low-phenylalanine diet can have on both the parents of PKU children and the children, themselves.

I thank my daughter, Leslie Barrick, for her tremendous help in the many battles that I had with my computer.

Finally, I thank my wife, Dr. Elaine Kaufman, for her patience in reading every section of the book for clarity. She pointed out many examples where I thought my explanations were clear but where she thought greater clarity was needed. She was invariably right.

TABLE OF CONTENTS

This is the story of the discovery, about 70 years ago, of phenylketonuria (PKU), a devastating genetic disease that spares the body but cripples the brain, causing severe mental retardation. It is the story of how the underlying cause of the disease was pinpointed to a metabolic defect in the liver and how, amazingly, this defect specifically blocks normal brain development. Finally, it is the story of how an accurate diagnosis and effective treatment for the disease were worked out.

Frank L. Lyman, M.D., in his technical book, *Phenylketonuria,* wrote that "the prevention of the mental retardation associated with PKU represents one of the great advances in medicine." And yet, outside of medical texts, it is a story that has never been told. *Overcoming a Bad Gene* tells this story.

CHAPTER 1
DISCOVERY OF PKU

Louis Pasteur, arguably the greatest French scientist who ever lived, was so successful at his research that people felt that he must be extraordinarily lucky. To which he responded, "In the field of experimentation chance favors only the prepared mind."

About 60 years ago in Norway, medical history was made because chance favored the prepared mind of Dr. Asbjörn Fölling. Not only was his mind well prepared, but he also happened to be in the right place at the right time.

The right time was 1934 and the right place was Oslo, Norway, where Fölling, a physician and biochemist, was Professor of Nutritional Medicine at the University School of Medicine. Chance came to him in the form of Mrs. Borgny Egeland., a young Norwegian mother of two severely mentally retarded children, Liv, a girl aged six and a half and Dag, a boy aged four. Mrs. Egeland. proved to be a major player in the medical drama that was about to unfold. Displaying extraordinary intuition and dogged persistence, she went from doctor to doctor trying unsuccessfully to find someone who might be able to treat her children. She did not give up because she seems to have sensed that to have one affected child might be a freakish, isolated misfortune, like being struck by lightning. But to have two affected children was uncanny, outrageous, like being struck by lightning *twice.*

That she was dealing with something unusual was underscored by her observation that a peculiar mousy odor, also described by others as the smell of horse stables, clung to both children no matter how often she bathed them. Was this just a coincidence or was there a connection between her children's illness and this peculiar odor?

To each doctor she visited, she recounted the same story about her children's early development. She told them that with Liv, everything seemed to be normal for the first few years. But when Liv was almost three years old and had still not said any words, Mrs Egeland and her husband, Harry, began to get worried. They began to wonder when she would start to talk and whether she was normal. When Mrs Egeland told their physician about her concerns, he was very reassuring.

There was nothing to worry about. He was sure that the little girl would talk later. Why, he, himself, did not start to talk until he was three years old. But, it turned out that Liv did not talk at three years of age or at four or even at five. In fact, she never learned to talk.

The Egeland's second child, Dag, was born three years after his sister. Like Liv, he was normal and alert for a few months but then seemed to get weaker and to lose interest in his surroundings. He was frailer than his sister was at that age.

Whereas Liv had walked when she was 16 months old, Dag was never even able to sit up by himself. Although he grew physically, he remained mentally like an infant. He died at six years of age.

In their desperate search for a physician who might be able to treat their two children, Mr and Mrs Egeland at last had a bit of good luck. It turned out that during his studies at the dental college, Harry Egeland, the father of the two retarded kids, had actually taken a course taught by Professor Fölling and had learned that he was doing research on metabolic diseases. It also turned out that Borgny's sister was acquainted with Dr Fölling. Borgny asked her sister if, when the occasion arose, she would tell Dr Fölling about her mentally retarded niece and nephew and, in particular, about their peculiar odor. Did he think there could be a connection between the odor and their mental retardation? Fölling had never heard of such a disease, but, his curiosity aroused, he agreed to examine the children. Mrs Egeland was asked to bring them together with a sample of their urine to the University Hospital.

After a routine clinical examination of the children, there was no doubt in Fölling's mind that they were both severely mentally retarded. The older sister could only say a few words. When she walked, her gait was stiff and twisted. She was also described as having a "whimsy" way of walking, often seeming to move around in a random manner. The younger brother appeared to be even more grossly retarded than his sister. He couldn't say a single word and was unable to walk or eat or drink on his own. He was not toilet-trained.

When Fölling went beyond a physical evaluation of the children, he quickly achieved a major breakthrough. As part of a thorough medical examination of his two small patients, he tested for the possibility of diabetes by carrying out a standard color assay for ketones, such as acetone, in their urine. The assay is simple. One merely adds a few drops of a 10% solution of ferric chloride to an acidified sample of urine. A positive test is indicated if the urine turns a purple or burgundy color. When Fölling carried out the test on the children's urine, however, to his great surprise the samples turned deep emerald green, a result that he had never seen before.

Before allowing himself to get too excited by this finding, he had to make certain that the green color was not due to some medicine that the children were taking. He also had to show that this result was reproducible. The mother brought in some new samples of urine with the assurance that the children had not been given any drugs or medicine before the urine samples had been collected. Again, Fölling observed the green color on adding ferric chloride to these fresh samples.

The results meant that some unknown substance in the urine was reacting with the ferric chloride to form a new colored compound. For Fölling, this must have been one of those exhilarating, once-in-a-lifetime eureka moments, the kind that every scientist longs for. The most likely explanation for his findings was that these mentally retarded children were forming and excreting an unusual substance that is not present in normal urine. The findings raised the possibility that this unknown compound could cause mental retardation. The results might even suggest a truly radical notion: that some error of metabolism resulting in the formation of this mysterious substance was the cause of the mental retardation. At a more immediate and practical level, the green color meant that he now had an easy way to follow the purification of this unknown compound, to track it while he tried to separate it from the thousands of other substances in the urine. Isolation of the pure compound was the essential first step before he could attempt to determine its structure.

Donning his chemist's hat, Fölling proceeded during the next few months to isolate the unknown substance from 20 liters of the children's urine. Once he had purified it, he was able, even using the limited armamentarium available to chemists at that time, to prove beyond any doubt that the abnormal compound was the keto acid called phenylpyruvic acid. His medical colleagues quickly dubbed it the "idiot acid".

Of the many questions sparked by Fölling's discovery of this new disease, one of the most urgent was whether these two kids were unique or were there other severely retarded patients who suffered from the same illness? To answer the question, he carried out the ferric chloride test on urine samples from 430 institutionalized mentally retarded patients. Eight of them were positive, giving an incidence in this group of about 2%. Later, more extensive surveys came up with a somewhat lower figure of 0.7%. These results showed that the new disease accounts for only a tiny fraction of the estimated 1% of the general population who suffer from mental retardation. Incidentally, phenylpyruvic acid has never been detected in urine samples from normal individuals.

There was an unexpected bonus that came from the survey of retarded patients that Fölling had carried out. Among those few that appeared to have PKU, there were 3 pairs of siblings. This pattern strongly suggested that he was dealing with a genetic or inherited disease, a conclusion that was fully validated by the subsequent work of others. Not only that, but more detailed analyses of the occurrence of the disease among relatives of the affected children (i.e., the "pedigree" of families with these children) showed that PKU is an autosomal recessive disease. In other words, the trait is not carried on a sex chromosome and it takes the inheritance of 2 "bad" genes, one from each parent, to get the disease.

In 1934, Fölling published in German, the popular language of science in those days, his classic paper describing his discovery of a new disease.The English translation of the title was "On excretion of phenylpyruvic acid in the urine as an anomaly of metabolism in connection with mental retardation." He suggested the name "oligophrenia phenylpyruvica" (Greek,*oligo*, little; *phrenia* ; mind). Since its discovery, many other names have been suggested, including phenylketonuria, abbreviated, PKU, the name that has caught on.

Pure phenylpyruvic acid is odorless. It was, therefore, not the compound responsible for the offensive smell that first brought Mrs. Egeland to Fölling to seek his help. Later, it was shown that this mousy odor was due to phenylacetic acid, a compound that is a close chemical relative of phenylpyruvic acid. Had Fölling followed his nose rather than a color test, he probably would have found phenylacetic acid instead of phenylpyruvic acid in the urine of these patients and the disease might well have been called "phenylaceturia".

With the identification of phenylpyruvic acid in the urine of these retarded children, Fölling had uncovered a critically important piece of what would prove to be a complicated puzzle. But what was the overall picture that this single piece fit into? Answering that question in the early 1930's turned out to be a daunting task- like trying to solve a jig-saw puzzle when most of the pieces were still missing.

The chemical structure of phenylpyruvic acid, itself, provided the first key to the solution of the puzzle. Indeed, it even afforded a dim glimpse of what the final picture might look like. The reason for this was because its structure is so similar to that of another naturally-occurring compound, phenylalanine, an amino acid that is a normal component of most proteins and is also present in blood and tissues. What was noteworthy about the relationship between these two compounds is that their structures are identical except that a single oxygen atom in phenylpyruvic acid (the one attached to the middle carbon (C) atom) is replaced in phenylalanine by a single nitrogen (N) atom. The structural relationship between these two compounds is shown in Fig. 1.

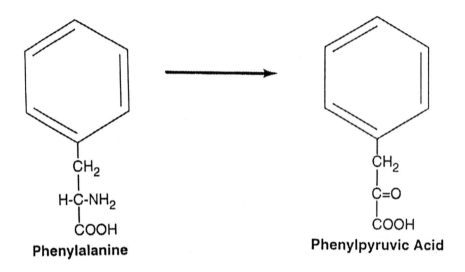

Phenylalanine   **Phenylpyruvic Acid**

Fig.1. The conversion of phenylalanine to phenylpyruvic acid

The significance of this striking similarity was that it meant that the two compounds were almost certainly metabolically related, metabolism being Nature's grand bazaar where miracles of chemical transformation are carried out. These miracles include the breakdown of the compounds in the food that we eat (proteins, fats, sugars) to simpler substances (with the concomitant release of energy) and the conversion of many of these simpler substances into complex molecules that the body needs. To go even further, the similarity between these two compounds made it highly likely that one of them could be derived from the other, i.e., they bore the same relationship to each other that a prune does to a plum or a raisin to a grape. Although this line of reasoning might even suggest that these two substances could be interconverted, the fact that phenylalanine is a widely distributed naturally-occurring amino acid, whereas phenylpyruvic acid is not normally found in man or other animals, indicated that the more likely sequence was that phenylpyruvic acid is formed from phenylalanine in a reaction that does not normally occur. In other words, these mentally retarded kids were suffering from a metabolic error, an abnormal conversion of phenylalanine to phenylpyruvic acid.

But if phenylalanine is not normally converted to the keto acid, what is it normally converted to? And was the metabolic error the cause of this new disease or a consequence of it? There were just too many pieces of the puzzle missing to answer these questions. In fact, it took four years before the next piece of the puzzle was uncovered. In 1938, perhaps guided by the structural similarity between phenylalanine and phenylpyruvic acid discussed above, Fölling made the stunning observation that PKU patients suffer from still another metabolic abnormality. In addition to excreting phenylpyruvic acid, the levels of the amino acid phenylalanine in their blood were markedly elevated, a condition now called hyperphenylalaninemia or HPA.

This discovery was a beacon, illuminating the outer boundaries of the puzzle. It was now possible to discern at least an outline of the nature of the metabolic disease that afflicted these retarded kids. Everything pointed to the conclusion that they suffered from a near-total block in the normal metabolism of phenylalanine, whatever that normal metabolism might be, a block that caused phenylalanine to pile up in their blood and their tissues. And with its normal metabolism short-circuited, the metabolism of phenylalanine was diverted or shunted into an abnormal route leading to the formation of an aberrant product- phenylpyruvic acid. If, in addition, it was assumed that phenylpyruvic acid is toxic to the immature brain, it would establish a direct connection between the metabolic block and the development of severe mental retardation. According to this scenario, phenylpyruvic acid was actually the "idiot acid". Such a scenario is reasonable. It is logical. But is it correct? Is phenylpyruvic acid a brain killer, a poison that can utterly devastate the brain in children with PKU? Unfortunately, the jury is still out on that question. Indeed, as we will see, there are other scenarios, perhaps even more appealing than the toxic metabolite theory. But before we can consider them, there are big chunks of the puzzle that still must be identified and put into place.

After Fölling's work (See Fig.2), the missing piece of the PKU puzzle that loomed the largest was the question of how phenylalanine is normally metabolized? Answering that question would be like killing two birds with one stone because that answer would not only fill in a gap in our understanding of how our bodies metabolize phenylalanine, but it would automatically identify the exact nature of the metabolic block in this disease.

Fig. 2. Dr. Asbjörn Fölling, biochemist and physician. This photograph was taken around the time of his discovery of phenylketonuria in 1934 (reproduced from Pediatrics 2000;105: 89-103).

CHAPTER 2
IDENTIFICATION OF THE METABOLIC BLOCK IN PKU

In 1934, when Fölling discovered PKU, what was known about how phenylalanine is normally metabolized was so fragmented that all attempts to put the few relevant facts together in a way that made sense had been stymied. A sign of just how muddled the field was at that time can be seen from the fact that more than a decade passed before the next great advance was made. The person most responsible for making that leap, for converting this chaos into order, was George Jervis, a European- trained physician-scientist who had received his medical degree from the University of Turin in Italy and two Ph.D. degrees, one in neurology from the University of Paris and a second one in psychology from the University of Milan. When he carried out the research that was to change medical history, he was the Director of Research at Letchworth Village, at the New York State Department of Mental Hygiene located in Thiells, New York. It is fair to say that if Fölling was the father of PKU, Jervis was the midwife.

In 1947, Jervis reported the results of a series of experiments that were straightforward in their execution and far-reaching in their significance. He showed that feeding

13

phenylalanine to animals and to normal humans led to a prompt rise in the amount of another amino acid, tyrosine, in their blood. This observation was coherent with results of an earlier German study, carried out at the turn of the century, that showed that pumping a solution of phenylalanine through a dog's liver led to the formation of extra tyrosine. Structurally, tyrosine, which is also widely distributed in proteins, resembles phenylalanine even more closely than does phenylpyruvic acid; it differs from phenylalanine only by the presence on the "phenyl" part of the molecule of linked oxygen and hydrogen atoms (called a hydroxyl group). This strikingly close structural similarity, when taken together with the speed of the increase in blood tyrosine following the administration of phenylalanine, strongly suggested that conversion to tyrosine is a step in the normal metabolism of phenylalanine (see Fig. 3).

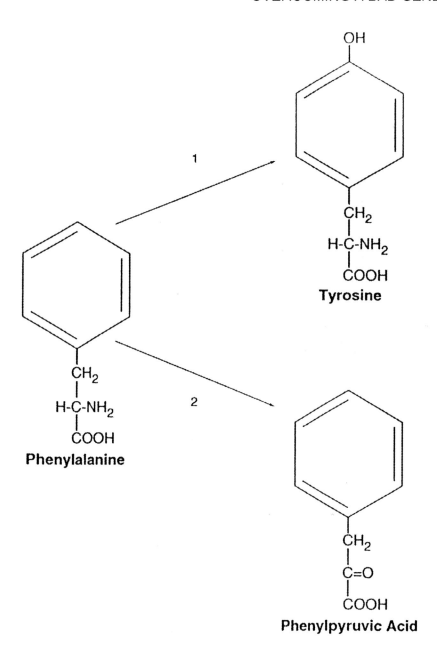

Fig. 3 The enzymatic conversion of phenylalanine to tyrosine
and to phenylpyruvic acid

As exciting as this observation was, it was topped by the results of the next experiment. For when Jervis gave phenylalanine to a group of PKU patients, there was absolutely no increase in their blood levels of tyrosine! With these relatively simple experiments, Jervis had solved the puzzle of the cause of PKU. On the basis of these results, he postulated that the metabolic error in PKU is an inability to convert phenylalanine to tyrosine. In other words, they were unable to hydroxylate (i.e., add a hydroxyl group) to the phenyl ring of phenylalanine. These findings of Jervis also went a long way toward proving that conversion to tyrosine is not only a step in the normal metabolism of phenylalanine, which can then be completely oxidized to carbon dioxide and water; it is an *essential* step. This last conclusion follows from the finding that when this step is blocked, as it is in PKU, the phenylalanine, which is derived from the breakdown of protein in the food that we eat, just sits there, piling up in the body. When its levels in the blood rise to 10 to 20 times above the normal level, a small portion of the phenylalanine is converted to compounds like phenylpyruvic acid, which, together with a fraction of the phenylalanine, itself, is excreted in the urine.

In 1953, six years after he had reported this pioneering work, Jervis published results of another study on PKU that clinched his earlier postulate that the metabolic lesion in PKU is the inability to convert phenylalanine to tyrosine. He managed

16

to obtain a snippet of liver tissue, i.e. a "biopsy", from two PKU patients, as well as from three patients who had died from other causes. He then added a small amount of homogenized liver from each patient to a buffered solution of phenylalanine and incubated the mixtures for one hour at close to room temperature. At the end of that time, he determined whether any extra tyrosine had been formed. The results were a dramatic confirmation of his earlier *in vivo* (Latin, *vivo*, in that which is alive) experiments. Whereas the control livers produced extra tyrosine, there was not a trace produced by either of the two PKU livers. From the results of these *in vitro* (Latin, *vitro* , glass) experiments, Jervis concluded that PKU patients suffered from a deficiency of the phenylalanine oxidizing enzyme system.

In carrying out this research, Jervis displayed a trait that characterizes all great experimental scientists: in contrast to most run-of-the mill scientists, who show a great instinct for the capillary of a research problem, he went right for the jugular.

Research carried out in the years subsequent to the work of Fölling and Jervis, filled in the details of the way phenylalanine is normally metabolized in humans. This work showed that the hydroxylation of phenylalanine plays a dual role in the metabolism of this amino acid. First, it provides the organism with its own supply of the amino acid tyrosine. As long as the hydroxylation reaction occurs, tyrosine is classified as a nonessential amino acid (i.e., a dietary source of tyrosine is not essential for normal growth). In the classical form of PKU, in which the hydroxylase is inactive or missing, tyrosine is an essential component of the diet, just as essential as any vitamin.

The second metabolic role for the hydroxylation reaction is that it is an obligatory step in the complete oxidation or combustion of phenylalanine to carbon dioxide and water. There are no alternative pathways by which the phenyl ring of this amino acid can be ruptured. Even in the complete absence of the hydroxylation reaction, phenylalanine can still be metabolized but the metabolism is restricted to the non-phenyl ring (also called the side-chain) part of the molecule. Fig. 4 outlines the main metabolic pathway for phenylalanine that is initiated by the hydroxylation step (reaction 1) and the secondary pathway that is restricted to the metabolism of the side chain of phenylalanine (reactions 2, 4, 5 and 6).

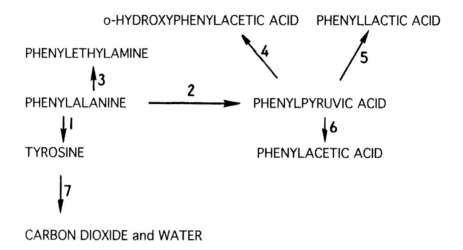

Fig. 4. The mammalian metabolism of phenylalanine. As can be seen, the hydroxylation reaction, the conversion to tyrosine (step 1), is the initial step in the pathway that leads to the complete metabolism to carbon dioxide and water (step7). Although for the sake of simplicitry this step is depicted as a single reaction, step 7 is actually the sum of more than six separate reactions. When reaction 1 is blocked, as it is in PKU, a reaction that is normally a minor one, conversion to phenylpyruvic acid, (reaction 2) becomes the major step. The latter compound can be further metabolized to compounds like phenylacetic acid (reaction 6), o-hydroxy-phenylacetic acid (reaction 4), and phenyllactic acid (reaction 5). It should be noted, however, that the pathway initiated by reaction 2 cannot lead to the complete oxidation of phenylalanine to carbon dioxide and water.

Together with the earlier findings of Fölling, Jervis' work firmly established that patients with PKU suffer from what at the time was a relatively new kind of disease, an inborn error of metabolism. This term was coined in 1908 by Sir Archibald Garrod, one of the towering figures in the field of inherited disorders, who also predicted, almost half a century before it had been proven, that these diseases were caused by the deficiency of an enzyme. Indeed, the proof by Jervis that the phenylalanine oxidizing system is defective in PKU patients provided one of the earliest experimental validations of Garrod's prediction.

It should be noted that although the work of Jervis and Fölling solved one of the mysteries that had been challenging scientists in this field ever since PKU had been discovered, their work did not provide any insight into what is perhaps an even deeper mystery: how is an elevated level of phenylalanine in blood translated into a profound deterioration in mental capacity?

The evidence that enzymes are the ultimate culprits in this kind of genetic disease focussed attention on these naturally-occurring substances. Enzymes are proteins, present in every living cell, from bacteria to plants to mammals, that can, as if by magic, catalyze or accelerate the transformation of one substance into another. They are the very wheels on which metabolism turns. To get some idea of the power of an enzyme, consider the digestion of starch, a reaction that takes place in our bodies every time we eat a piece of bread. In the absence of an appropriate enzyme and under the mild conditions that prevail in our tissues, the digestion of starch (ultimately to sugar) might take hundreds of years. By contrast, the digestive enzyme amylase, which is present in saliva and the pancreas, can do the job in a matter of minutes. Without resorting to strong acids or alkalies and elevated temperatures, not even the best chemists in the world can come close to matching what this enzyme can readily do at body temperatures under mild conditions. In the absence of enzymes, the vital metabolic transformations would take place far too slowly to sustain life.

When Jervis succeeded in showing, both in *vivo* and in *vitro*, that PKU patients are unable to convert phenylalanine to tyrosine and that a defective phenylalanine hydroxylase molecule was the cause of the problem, there was no reason to doubt that his work had closed this chapter in the PKU story. Subsequent research, however, unexpectedly reopened it.

In the mid-1950's, while working at the National Institutes of Health (NIH), I became interested in learning how the normal phenylalanine hydroxylating enzyme system functions and how genetic defects (i.e., mutations) in this system can cause PKU.

During the early stages of this work, I had an idea about how excess phenylalanine might damage the brain. I thought the idea could be tested by carrying out a relatively simple

21

nutritional study on a seriously retarded PKU patient. Fortunately, I was able to locate a suitable patient, a handsome 5 year old boy who was living in the Washington area.

When the patient was brought to NIH by his parents, I proceeded to explain to them the rationale for the study that I planned to carry out. During this meeting, I reached out spontaneously to the boy and lifted him up to my lap, as I often did with my own son, who was just about the same age as the patient. The difference, however, between holding my son and holding this young boy startled me. Unlike the pleasant experience of holding my son, or for that matter, any other normal youngster of that age, trying to hang on to this PKU boy was like trying to control a tightly coiled spring. When the boy's mother noticed my discomfort, she nodded knowingly and said that even she had trouble holding him. He finally squirmed off my lap and spent the rest of the session sitting in a corner on the floor.

I was awed by the experience. Holding that child on my lap made me realize, as few other experiences could have, what terrible devastation had been wrought on this child's brain by the defective functioning of only a single enzyme among the more than 2000 enzymes in our bodies.

Although it turned out that there were too many obstacles to carrying out the study I had planned to do on this young patient, my brief hands-on encounter with him continued to haunt me. The experience further increased my interest in learning more about the phenylalanine hydroxylating system-PKU connection.

Being trained as an enzymologist, I took it as an article of faith that learning more about the enzyme system would teach us something important about this brain-crippling disease. Ideally, the best way to gain some insight into how a defective human enzyme molecule, already known to be present in liver, can lead to a defective brain would be to study this particular enzyme in human liver. Getting a large piece of fresh human liver for research purposes in civilized countries, however, has always been close to impossible. Given enough effort, one can usually manage to get a piece of human liver at autopsy, but trying to get an accurate picture of an enzyme's properties from dead tissue would be a very iffy proposition. Despite Dr. Frankenstein's success in constructing his monster from a "lifeless thing" through the simple expedient of zapping it with a few bolts of lightning, as depicted in the movie, in the real world this approach would be doomed. In the real world, autopsy tissue is generally unsuitable as a source of enzymes because these labile proteins usually deteriorate and lose their enzymatic activity rapidly after death.

Fortunately, this dilemma can often be side-stepped by taking advantage of one of the invaluable guiding principles of enzymology- the concept of the "Unity of Nature". This generalization teaches that the vast majority of physiological and metabolic processes are essentially identical or at least very similar in humans and animals and even in bacteria. This principle was a priceless aid in a study of the phenylalanine oxidizing system. Without it, any study of this system and its connection with PKU would be forced to focus on the enzyme in human liver. On the other hand, guided by the unity principle meant that the enzyme system could be explored in rat liver with every assurance that any findings would also be true for the human liver system.

Building on the work of others, which is almost always the way that science advances, I showed that in rat liver the phenylalanine hydroxylating system is surprisingly complex, made up of two enzymes and a new non-protein coenzyme. Coenzymes are small mobile molecules that are essential for the activity of certain enzymes. Until these enzymes combine with their specific coenzymes, these enzymes are inert, just as inert as an internal combustion engine would be without a sparkplug. I showed that the coenzyme in this case is one of the class of compounds called a "pterin", specifically, tetrahydrobiopterin, a distant relative of the vitamin folic acid . Tetrahydrobiopterin bears a substituent in the 6 position of the pterin ring and is occasionally called simply 6-biopterin. I also showed that one of the enzymes in the hydroxylating system is phenylalanine hydroxylase, the enzyme that actually puts the hydroxyl group on phenylalanine and converts it to tyrosine and the other one is dihydropteridine reductase (DHPR) that maintains tetrahydrobiopterin in the active reduced form. Each

24

of these three components was absolutely required for the hydroxylation of phenylalanine. This meant that the lack of any one of them could cause PKU.

My wildest hope was that tetrahydrobiopterin would prove to be the offender because that would mean that PKU could be treated simply by giving this substance to the patients, whereas giving them a missing enzyme was not then, or even now, a very feasible therapy. Here was a striking example of how the light generated by basic research might illuminate a murky clinical area and forever change it.

Sometime in 1954, after contacting dozens of clinicians who I knew were treating PKU patients, I finally located one who was willing to try to obtain liver biopsy samples from several patients, as well as from a few non-PKU controls. Within two weeks, the pea-size frozen samples arrived packed in dry-ice. Using quantitative assays that I had developed, I showed that all of the PKU liver samples had normal amounts of tetrahydrobiopterin as well as the reductase but were totally devoid of phenylalanine hydroxylase activity. This was a good news-bad news result. The good news was that at last, almost 20 years after Fölling had discovered PKU, the missing enzyme had been unambiguously identified: phenylalanine hydroxylase. The bad news was that it was a missing enzyme rather than a missing coenzyme. Twenty years later, as we will see, this bad news was superseded by an unexpected and dramatic development.

If one can dare say that there is anything fortunate about a disease as dreadful as PKU, it is that it is extremely rare. As already mentioned, it accounts for about 1 % of the population who suffer from all forms of mental retardation. In the U.S., the incidence of PKU is roughly 1 in 12,000. The frequency, however, varies widely among different ethnic groups, being only 1 in 200,000 in Japan but as high as 1 in 5000 in Ireland. It is also high in Turkey, West Scotland and amongst the Slavs. On the other hand, it is extremely rare in Afro-Americans and Ashkenazi Jews (but high in Yemenite Jews).

Although the disease is rare, the number of cases add up: it has been estimated that, currently, there are between 15000 and 20,000 PKU patients in the U.S. This number remains fairly constant because the pool of patients is replenished with about 300 new cases being added each year- almost one a day.

The vast majority of PKU babies are born to parents who are both heterozygotes or carriers of the disease but who are otherwise healthy with normal IQs. They harbor 2 different kinds of genes or alleles (an "allele" is one of two or more alternate forms of a gene) . Each of the 2 alleles dictates the synthesis of 2 different forms of phenylalanine hydroxylase, 1 normal fully active form of the enzyme, controlled by the normal gene and the other a mutant form with little or no hydroxylase activity, controlled by the mutated gene.

When it comes to the genetic makeup of the offspring born to such heterozygous parents, these babies are unwitting players in a high-stake genetic crap game. On average, one quarter of them will be normal, having inherited 2 normal hydroxylase genes, one from each parent, one quarter will have PKU, having inherited 2 mutated genes and one-half will be heterozygotes, just like their parents, having inherited a normal gene and a mutated gene, one from each parent.

CHAPTER 3

GENETIC DISEASES: NORMAL GENES, MUTATED GENES, HETEROZYGOTES

What is a mutated gene? For that matter, what is a gene? As defined in the dictionary, a gene is "a unit of inheritance", which means that genes carry all of the genetic information needed to fashion a whole organism, down to the smallest detail- be it a moth or a monkey or a man, including, for man, the color of the eyes and hair to the shape of the nose, to the length of the earlobes, to the susceptibility to a multitude of serious diseases such as certain types of colon cancer, diabetes, muscular dystrophy and hypercholesterolemia. The 30,000 to 40,000 genes, mostly located on the chromosomes in the nuclei of our cells, are akin to an architect's blueprints that describe every detail about how a house is to be built. And, just as a well- placed stick of dynamite can topple a magnificent building, so a single mutated gene can topple the magnificent entity called man. Whereas it used to be thought that our destinies are written in our stars, we now know that they are written largely in our genes.

Genes function in our cells by directing the synthesis of proteins, including enzymes such as phenylalanine hydroxylase. The mother's egg contains genes distributed on

23 nuclear chromosomes. The father's sperm also contains copies of these genes on his 23 chromosomes. At conception, one set of these chromosomes from the mother combines with the set from the father to give the offspring a full complement of 23 pairs of chromosomes or a total of 46 chromosomes. Even if two enzymes function as part of a metabolic unit, as do phenylalanine hydroxylase and dihydopteridine reductase, the genes that direct the synthesis of these enzymes are not necessarily located close together or even on the same chromosome. In fact, the hydroxylase gene is located on chromosome 12, whereas the gene for dihydropteridine reductase is on chromosome 4.

At the chemical level, genes are made up of long molecules of deoxyribonucleic acid, abbreviated DNA, often called the "blueprint of life". The building blocks of the DNA are 4 different kinds of smaller molecules collectively called "nucleotides". Nucleotides are composed of even smaller molecules called "bases"- purines or pyrimidines-, each combined with a molecule of sugar and a phosphate group. In mammalian DNA there are 2 different purines, adenine (A) and guanine (G) and 2 different pyrimidines, cytosine (C) and thymine (T). In the DNA molecule these bases occur as pairs, in which the different bases are loosely connected to another base in a non- random pattern. Adenine is loosely connected or paired with thymine and guanine is paired with cytosine. The nucleotides are strung out on a thread of DNA like beads on a necklace. It has been estimated that all human cells (except sex cells-sperm and egg cells- and red blood cells that do not have nuclei) contain about 6 feet of DNA molecules. The DNA is believed to be made up of 3 billion base pairs. Scientists have further estimated that if these base pairs were printed out, they would fill more than 1,000 Manhattan telephone directories.

The information contained in the DNA molecule constitutes what is known as the genetic code. The information in the code is determined by the unique sequence of these nucleotides that, in turn, dictates the unique sequence of amino acids in a particular protein. The fundamental unit of the code is a cluster or triplet of 3 adjacent nucleotides known technically as a "codon". It turns out that for a language based on 4 letters, used 3 at a time, there are 64 (4 x 4 x4 = 64) different possible combinations of these letters. Since proteins are constructed of only 20 different amino acids, the 4 letter alphabet of DNA is more than sufficient to specify the unique sequence of amino acids in all of the known proteins. It is at the level of DNA and proteins that we see the astonishing resemblance between man and all the other animals on this planet. It has been shown, for instance, that humans and chimpanzees have in common about 98.5% of their DNA. This striking resemblance has led the physiologist Jared Diamond to call man "the third chimpanzee". As for proteins, it has been found that the resemblance between the structure of certain of them in man and in animals is awesome. Given the overlap in their DNA structures, for example, it may come as no surprise that the amino acid sequence of certain proteins in man, such as one called cytochrome c, is identical with its counterpart in chimps. But the similarity extends even to those animals who do not look anything like us, such as turtles and rattlesnakes, whose cytochrome c shares 85% of their amino acids with that of man.

The way the language of the DNA code is deciphered can be illustrated by designating the 4 different nucleotides or letters in the code simply by the abbreviations of their purines and pyrimidines, i.e. A,G,C, and T. With that designation, it has been shown that the triplet AAA specifies the location of the amino acid lysine in the protein being synthesized and the triplet GAA has been shown to specify the position of the amino acid glutamic acid, and so forth. The sequence of nucleotides in this stretch of DNA would then be AAAGAA. and the corresponding sequence of amino acids in the newly synthesized protein would be lysine-glutamic acid.

Inherited mutations can arise on exposure of an organism to some kind of noxious factor in the environment, such as X-rays, cosmic rays, radon or certain industrial chemicals. Exposure to these agents can alter the structure of the DNA within the germ cells, i.e., the sperm or the egg, that are passed on to the next generation at the time of conception. The result may be an inherited metabolic disease such as PKU. On the other hand, exposure of your non-germ cells, such as your skin cells, to a mutagenic agent may damage you, rather than your offspring, by increasing your chances of getting cancer.

One of the more common structural changes caused by mutagenic agents is the conversion of a purine or pyrimidine in one nucleotide in the DNA sequence to a different one, e.g. the conversion A in the first position of the codon AAA, which codes for the amino acid lysine to another one, e.g. G, to give the altered codon GAA, which codes for the amino acid glutamic acid, resulting in a mutated gene. If the newly synthesized protein happens to be an enzyme, such a switch often results in a useless, inactive enzyme. With respect to phenylalanine hydroxylase, two mutant alleles of this gene usually differ in only a single amino acid substitution at the site of a mutation.

A useful shorthand method of designating the different allelic forms of the hydroxylase (or any other enzyme) uses the name of the mutation that they harbor. For instance, a mutation that leads to the replacement of the amino acid arginine (symbol R) with the amino acid tryptophan (symbol W) at amino acid residue 413 of the enzyme would, by this method, be designated R413W. This happens to be the most frequent PKU-associated mutation in the hydroxylase molecule in Europeans.

The first experimental proof that mutated genes actually do their dirty work to the host organism by changing the structure of proteins was provided in 1949 by one of America's preeminent chemists, Linus Pauling, a future Nobel Prize winner, working at the California Institute of Technology. Pauling had already made important contributions to our understanding of the structure of hemoglobin, including details about how this vital protein carries out its physiological function of combining with oxygen and carrying it to tissues, when he first heard about a terrible genetic disease called sickle cell anemia that might involve hemoglobin. One of the most common inherited ailments, it affects mainly people of African descent and can cause life-threatening medical emergencies or chronic illness. The possibility that this type of anemia might involve hemoglobin fell on fertile ground because Pauling's earlier fundamental basic studies on this protein had already prepared his mind to receive this new information.

His interest in sickle cell anemia was especially piqued when he learned that only the red cells in venous or deoxygenated blood from these patients assume a sickle shape and that they regain their normal shape in the arteries where the blood is oxygenated. The twisted red cells become sticky and clump together, interfering with the flow of blood and the delivery of oxygen to some parts of the body. The twisted red cells are also rapidly removed from the circulation, causing anemia.

The finding that the sickling process depends on oxygen suggested to Pauling that hemoglobin was involved in this process since it was known that it was hemoglobin in red cells that combines with oxygen. In 1949, Pauling and his colleagues published a seminal paper in the journal *Science* entitled "Sickle Cell Anemia, a Molecular Disease". In this paper, they showed that at least one property of hemoglobin from patients with this disease differs from that of normal hemoglobin. The distinguishing feature of the patient's hemoglobin was its electrical charge. The difference was sufficient to permit its separation from normal hemoglobin when both proteins were subjected to an electric current in a technique called electrophoresis. This altered property meant that the *structure* of the patient's hemoglobin was different from normal hemoglobin. Using the same technique, Pauling and his coworkers showed that the blood of carriers of sickle cell anemia contained about a 50-50 mixture of normal and sickle cell hemoglobin.

This work demonstrated that genes could indeed alter the structure and the properties of proteins. Since an altered molecule- in this case the molecule was hemoglobin- was the cause of the disease, Pauling coined the term "molecular disease" to describe this kind of ailment. In 1956, he further elaborated this notion with the following definition: "A disease of this sort, caused by molecules of abnormal structure present in the patient in place of the molecules of normal structure that are present in normal human beings, is called a molecular disease." Pauling extended the concept of molecular disease to include some kinds of mental deficiency. Ralph Gerard, a neuroscientist, carried this notion still further when he speculated, to paraphrase his statement, that behind every twisted thought there was a twisted molecule.

In 1956, in discussing the concept of molecular diseases, Pauling specifically suggested that PKU falls into this category. He speculated that in the future artificial enzymes would be used to treat such diseases. He envisioned the time when a synthetic catalyst that can oxidize phenylalanine to tyrosine could be attached to a reticular framework inside of a small open-ended polyethylene tube which could be permanently placed in an artery of a PKU baby where it could cleanse the blood of excess phenylalanine. Although nobody has yet succeeded in making an artificial phenylalanine hydroxylase, we will discuss later an experimental therapy for PKU that uses an alternate phenylalanine-metabolizing enzyme that, when administered to a PKU animal was able to decrease the animal's hyperphenylalaninemia.

The only part of the sickle cell anemia puzzle that eluded Pauling and his coworkers in 1949 was the identity of the structural change in the sickle cell hemoglobin molecule.. The missing piece was provided in 1957 by Vernon Ingram at the University of Cambridge who found that of the nearly 300 amino acid residues of normal hemoglobin, only one is different: one of the glutamic acid residues of normal hemoglobin is replaced by a valine residue in sickle cell hemoglobin. Since glutamic acid is a negatively charged molecule, whereas valine is an uncharged molecule, this structural change could explain why normal hemoglobin can be separated from sickle cell hemoglobin during electrophoresis, a technique that separates proteins from one another on the basis of their electrical charges.

With respect to PKU, in 1986 Savio Woo and his colleagues, working at the time at Baylor College of Medicine in Houston, accomplished the equivalent step with phenylalanine hydroxylase that Pauling and Ingram had accomplished with sickle cell hemoglobin when the Texas workers identified for the first time a mutation in the phenylalanine hydroxylase gene from a PKU patient that leads to an inactive enzyme and thereby causes the disease. Since then, more than 400 different mutations have been identified. In view of this large number of mutations, it is not surprising that most PKU patients have been shown to harbor two different kinds of mutations at the same locus. These patients are therefore referred to as "compound heterozygotes."

Returning to the heterozygous parents of PKU babies with their one good gene and one "PKU gene", why don't they have PKU? Or, if not the full-blown disease, why don't they have at least a mild form of it? What protects these carriers from the onslaught of the disease? The answer is that they are protected for the same reason that we can get along quite well with only one of our 2 kidneys, because 2 kidneys provide us with something of a cushion, with more kidney capacity than we normally need. In this respect, Nature is like the ultimate worrier- the man who wears a belt in addition to wearing suspenders to keep his trousers up.

Nature is just as generous when it comes to the amount of phenylalanine hydroxylase that she provides us with. As is true for most enzymes, our normal complement of hydroxylase is more than we usually need. For this reason, even though PKU heterozygotes have only about half the normal amount of hydroxylase, they do not seem to suffer any ill effects from this deficit. Their blood phenylalanine levels, which are mainly controlled by the amount of phenylalanine hydroxylase in their livers, for example, are only about 1.5-fold (50%) higher than the levels in normal individuals who have a full complement of the enzyme. This is in striking contrast to the 20-to 30-fold increase in blood levels of phenylalanine seen in PKU patients.

As already noted, PKU is a recessive genetic disease, which means that a baby must inherit 2 mutant genes to have this condition. It is recessive because of Nature's bounty, because we have something of an excess of the hydroxylase. Without such an excess, PKU would likely be classified as an autosomal dominant genetic disease, such as Huntington Disease, the disease that struck down the singer Woodie Guthrie. The hallmark of such diseases is that the heterozygotes, who have one normal and one mutant gene, can be almost as sick as homozygotes. In contrast to PKU heterozygotes, their normal gene doesn't seem to protect them, possibly because that particular gene provides no "cushion". An unwelcome feature of recessive genetic diseases is that even when the homozygous condition is rare, as is true for PKU, the incidence of heterozygotes for the disease can be high. For PKU, it is a startling 2% of the population. It shouldn't be surprising, therefore, that some geneticists have estimated that 2 out of 3 people are carriers of at least one serious recessive disease. Whenever I give a lecture on PKU to a group that numbers between 50 and 100, my announcement that there are probably 1 or 2 PKU heterozygotes in the audience usually grabs their attention.

In the U.S., an incidence of 2% translates to close to 5,000,000 people, which is about equal to the population of the entire state of Maryland. The overwhelming majority of these carriers are totally unaware that every cell in their bodies harbors one copy of the potentially harmful PKU gene. Without knowing their special genetic status, these millions of people don't realize that their chances of marrying another carrier are dangerously high- about 1 in 50. The scary thing about this kind of unlucky encounter is that the odds that such a couple will have a PKU baby are 1 in 4!

As mentioned previously, the incidence of PKU in some regions like Ireland and the western part of Scotland is relatively high. Although the reason for this is not known, it has raised the possibility that in the high incidence countries there must be some advantage to being a PKU heterozygote. A clue to what this advantage might be was uncovered by L.I.Woolf. He reported in 1994 the interesting finding that despite the fact that the number of pregnancies in PKU and control families were equal, in PKU families fewer of the pregnancies ended in spontaneous abortion. As a result, pregnant PKU heterozygotes would have 7.4 % more live-born offspring than women who are not carriers of the PKU gene and would therefore have 7.4% more descendants. One half of these children would carry the gene for PKU. In this way, the proportion of PKU heterozygotes would increase in the population and would lead ultimately to a higher incidence of the disease. These statistics indicated that the mother's heterozygosity was capable of protecting the fetus from some peril lurking in the environment in countries such as Ireland and West Scotland and probably in others as well.

What is this environmental hazard? Woolf proposed that the culprit is a naturally- occurring derivative of phenylalanine, ochratoxin A. This toxic compound is commonly produced by several species of fungi, *Penicillium verucossum* and *Aspergillus ochraceus,* that often infest stored grains such as corn, barley, oats, wheat, rice and, in addition, soya beans and coffee beans. Not surprisingly, the toxin is also found in tissues of farm animals that are fed stored grains, as well as in tissues of humans (blood serum, milk and kidney).

Ochratoxin A is nasty stuff. It has been shown to be extremely toxic to the kidneys in all animal species tested with the exception of mature ruminants. In rats and mice it is carcinogenic. Of special relevance to its possible role in conferring an advantage to PKU heterozygotes, is the observation that when the compound was given to pregnant mice, it produced a spectrum of effects ranging from fetal malformations to prenatal mortality. Of even greater relevance is the finding that in mice its toxicity to kidneys could be prevented when the toxin was given concurrently with phenylalanine. In view of the fact that ochratoxin A is a phenylalanine derivative, the finding that phenylalanine can compete with it and diminish its toxicity is not unexpected.

In trying to find a connection between ochratoxin A toxicity and the reproductive advantage of PKU heterozygotes the last observation appeared to provide a crucial missing link. Since the most obvious biochemical difference between normals and PKU heterozygotes is that the blood and tissue levels of phenylalanine are about 50 % higher in the latter group, the ability of phenylalanine to overcome at least one manifestation of the toxicity of the toxin shows one way in which heterozygosity can decrease the incidence of spontaneous abortions: the higher phenylalanine levels in the heterozygous mother may be able to protect her fetus from the bad effects of ochratoxin A.

Although Woolf's theory is provocative, evidence in support of it is scanty. For example, the theory would be more convincing if it could be shown in experimental animals that phenylalanine can actually protect the developing fetus from ochratoxin A-induced damage and death.

To put the ochratoxin theory in perspective, it is important to mention that the general concept of heterozygote advantage in certain genetic diseases is on very firm footing. The most notable example of this phenomenon comes from studies of sickle cell anemia where it has been found that carriers of the sickle trait, i.e, heterozygotes, are protected against malaria. It has been suggested that the mechanism of the protection is that the de-oxygenated red cell in heterozygotes can sickle and thereby crush the malaria parasite that transmits the disease (the parasite is delivered into the blood of the host by a bite from a heavily infected anopheles mosquito). In this manner, the host would be afforded some protection against contracting malaria.

For anyone interested in learning about some fundamental aspects of how the human body works, PKU is a great teacher. One of the most far-reaching lessons is that the different organs in the body continuously communicate with each other and that this dialogue is crucial for life. To paraphrase what the English poet John Donne said about "man", PKU teaches us that no organ is an island unto itself. We learn this lesson from PKU because the metabolic disturbance that underlies the disease is the result of a mutational hit taken by phenylalanine hydroxylase, an enzyme that occurs only in liver, whereas the only organ that is totally devastated is the brain. As will be discussed later, a successful treatment for this disease must be able to stop this particular kind of noxious communication.

CHAPTER 4
SYMPTOMS OF PKU

The most glaring manifestation of this brain damage is the profound mental retardation seen in most untreated patients. The majority have IQs less than 20. Far more telling than the numerical value of their IQs, however, which is notoriously difficult to measure accurately when the impairment reaches this catastrophic level, is a narrative description of their limitations. It is hard to improve upon the description of severe mental retardation such as that seen in most untreated PKU patients than the description of idiocy (the technical term used in medical or psychology books to describe people with this level of severe mental deficiency) given by the Encyclopedia Britannica: "a degree of mental deficiency so severe that the individual is incapable of attending to his personal needs (eating, elimination, dressing), protecting himself from ordinary dangers, mastering any useful command of speech or performing useful work". This picture jibes pretty well with Fölling's description of his first patients.

In addition to mental retardation, these untreated patients show signs of extensive neurological disease. The great majority have abnormal EEG patterns. Abnormalities of the encephalogram are much more frequent than clinical manifestations of convulsive disorder. Nonetheless, many- about 25%- do have a history of seizures, usually starting between 6 and 18 months of age. A large fraction of them have tremors, which are especially evident when the hands are stretched out without support. Watching these patients, one is struck by their almost continuous restless, purposeless, repetitive body movements, including aimless fiddling of the hands and fingers in what has been described as "pill-rolling" motion, technically called athetosis. As mentioned previously, even the mother of the first 2 patients who were examined by Fölling mentioned that the way her children walked was not normal. In general, the gait of untreated patients is stiff, consisting of short steps with the body bent rigidly forward. Many of them sit in a characteristic crossed leg position, called "schneidersitz" (from the German word describing the way a tailor sits when sewing). Adding to the picture of abnormal brain development in these patients, microcephaly is present in more than half of them and appears to be more pronounced in those with the greatest mental defect. Head circumference averages nearly an inch smaller than normal. Not surprisingly, delayed speech is not uncommon in children with PKU.

Even if one subtracts the severe mental retardation from the catalog of symptoms that characterize PKU patients, what remains adds up to a devastating illness. In fact, disturbed behavior, especially amongst those who are not so severely retarded that they display no "behavior", has been considered to be one of the most salient features of the disease. Just how striking a feature it is can be seen from the frequency with which PKU used to be diagnosed as a purely behavioral abnormality. For example, in the mid-1950's it was not that unusual for the majority of the patients who were ultimately diagnosed as having PKU to be admitted to institutions not because of their mental retardation but because of their grossly disturbed behavior. The disturbances run the whole gamut, including hyperactivity, episodes of sudden and unprovoked screaming and severe temper tantrums, as well as outbursts of violent behavior. Superimposed on such episodes, the patients were withdrawn, often failing to respond to human contact. This whole pattern of behavior has been interpreted by some doctors as psychotic. PKU patients were also frequently misdiagnosed as suffering from infantile autism.

Here is another valuable lesson that we learn from PKU, namely that the symptoms of psychosis, one of the most mysterious and scary illnesses known to man, can be provoked by a disturbance in the metabolism of a normal component of our bodies, the amino acid phenylalanine. It is one of the few instances where a clearly defined enzyme deficiency has been shown to mimic a psychiatric disorder. But what is the convoluted thread that connects these two seemingly disparate diseases i.e. Psychosis and PKU? In a later section we will discuss a possible link.

The behavior of phenylketonurics has been concisely summed up by a doctor who has studied them: "None could be described as friendly, placid or happy." As Pearl Buck, the writer, has poignantly noted, "they are living yet not alive". It has also been pointed out that the personality of these patients contrasts sharply with the outgoing friendliness of another group of mentally retarded patients, namely, those suffering from Down Syndrome.

Beyond their mental, neurological and behavioral defects, the general appearance of phenylketonurics belies the common depiction of morons and idiots, especially the one favored by Hollywood, as ugly oafs with misshapen bodies and distorted features. Although they frequently have eczema, phenylketonuric babies and young children are often singled out for their beauty, most of them having blond hair, light blue eyes and fair skin.

To most casual observers, these characteristics would probably add to the picture of a handsome baby; to the informed physician, it could be an early confirmation of a medical problem, especially if it goes along with signs of slow development. Deficient pigmentation is one of the common (but not invariant) hallmarks of PKU. Whereas it may be common for Norwegian couples to have babies with blond hair and blue eyes, the birth of a platinum-blond baby to dark-skinned Spanish or Sicilian families, as has happened occasionally when the babies are phenylketonuric, is noteworhy.

In addition to dilution of pigment with fair skin and hair, PKU patients display other extraneural signs and symptoms. The historically important "mousy" odor has already been discussed. Frequent vomiting and poor growth (as measured by height and weight indices) in untreated patients has been reported. About 25% of PKU patients have some form of eczema, with scleroderma-like lesions. (Scleroderma is a thickening and hardening of the skin with pigmented patches). It should be noted that whereas it is likely that all PKU patients will have some of the characteristic signs and symptoms, not every patient will have every one of them.

As the use of Fölling's urine test for PKU became more widespread, the disease began to be diagnosed more frequently during early infancy. By the mid-fifties, a totally unexpected feature of PKU, one with momentous implications, started to emerge: these babies appeared to be normal at birth and only became severely defective during the first six to twelve months of life. To the parents of these children who watched helplessly while their seemingly normal babies were being transformed into grossly retarded ones, this deterioration must have been the cruelest of ironies.

But there was a bright side to this gloomy picture. Finding a way to *prevent* the brain damage would almost surely be more feasible than finding a way to *reverse* it. Even a headache is usually easier to deal with if it is treated before the pain gets too bad.

As early as 1951, shortly after Jervis had shown that PKU is caused by a block in the normal hydroxylation of phenylalanine, which leads to the accumulation of the amino acid, some doctors wondered whether the disease could be treated with a diet low in phenylalanine. Although the logistics of carrying out such a treatment might be a nightmare, the idea was eminently reasonable. There was little reason to doubt that either phenylalanine or a metabolic product formed from it was toxic to the developing brain and the cause of the mental retardation. And if this was so, limiting the intake of phenylalanine should help. Furthermore, since it was known that one of the most critical features of brain development i.e., the formation of connections between neurons that allows them to "talk" to each other, is nearly complete by the 6th or 8th year after birth, it seemed likely that the treatment wouldn't have to last beyond this age.

CHAPTER 5
TREATMENT OF PKU

In the period between 1953 and 1955, several groups of clinical scientists, including Horst Bickel working in Birmingham, England (and later in Germany), Louis I.Woolf in London, and Marvin Armstrong in Salt Lake City, treated small numbers of phenylketonuric patients with a low-phenylalanine diet. Bickel was actually the first one who had the courage and the means to weave his way through this potential minefield. The right path could lead to a way to prevent the horrendous effects of the disease but a misstep could lead to disaster. Phenylalanine, after all, was known to be essential for life and the contemplated treatment involved markedly cutting back on the amount of this vital nutrient that these babies would be fed. In fact, there were even some careless suggestions in favor of feeding a "phenylalanine-free diet". This surely would have been a treatment far worse than the disease, since it would, sooner or later, kill the patient.

But where to start? You couldn't go down to your local supermarket and simply buy a box of low-phenylalanine food; you had to prepare it yourself, and large amounts of it. The approach that Bickel used was to start with casein, a cheap and readily available milk protein, and boil it with acid to break it down to its constituent amino acids (the resulting product known as a "casein hydrolysate.") The hydrolysate was then mixed with charcoal which is noted for its ability to combine avidly with phenylalanine, tyrosine and a few other amino acids. The charcoal with these few amino acids sticking to it was then separated from the rest of the casein hydrolysate, leaving a mixture of all of the other 17 or so amino acids, which was the starting point for formulating a low phenylalanine diet.

The original amino acid composition of the casein was reconstituted by adding back any of the amino acids that had been removed by the charcoal treatment, such as tyrosine, and then topping it off with a small amount of phenylalanine. The amount of phenylalanine was adjusted so that it would provide the babies with about 200 to 400 milligrams (0.007 to 0.014 ounces) of the amino acid each day, roughly the amount in 7 ounces of cow's milk or 10 ounces of human milk. This was enough phenylalanine for a baby's normal growth but far less than the amount that would be fed to a normal baby. Finally, to convert this mixture of amino acids into something resembling the diet that babies usually eat, foods very low in protein (therefore low in phenylalanine) , such as vegetables, fruits, butter or cream and sugar, were added back to provide carbohydrates, fats, minerals and vitamins.

Armstrong took a different route toward the goal of a low-phenylalanine diet. Rather than starting with a casein hydrolysate, he mixed together all of the needed amino acids in

50

amounts approximating those found in a protein like casein with the exception of phenylalanine, which was added back in limited amounts.

In the first brave experiments with this kind of dietary treatment, 19 phenylketonuric infants and young children, ranging in ages from 5 weeks to 8 years, were fed a low-phenylalanine diet for periods up to two and a half years. The earliest reports were tantalizing. In one of these publications, Bickel and his colleagues described the treatment of a two-year old PKU patient of Irish decent who was severely retarded. She could neither sit nor stand and seemed to take no interest in her surroundings. In December, 1951, when the girl was two years, two months old, the Bickel group initiated her treatment with the low phenylalanine diet that they had prepared. Almost all of the symptoms improved: the light hair darkened, the bad smell disappeared concomitant with the end of the excretion of the offending phenylalanine metabolites, most seizures stopped, there was less disruptive behavior, she began to develop good interaction with her mother and the nurses. In general , her awareness increased and, as expected, her blood phenylalanine levels decreased to the normal range because much less of the amino acid was being eaten. But her I.Q. values barely budged. These results were like looking at a glass that could be viewed as either half full or half empty.

Fortunately, this period of uncertainty did not last long. As younger and younger infants were treated, the verdict on the effectiveness of the diet was loud and clear. The low-phenylalanine diet, when started early enough, i.e., within the first few weeks of life, seemed to be able to prevent all signs of brain damage in these children, including the mental defect. The diet led to the normalization of their I.Q.'s. The treatment could thwart their cruel genetic destiny, could halt their relentless spiral down toward the dungeon of severe mental retardation (see Fig. 5).

Fig. 5. Photograph showing the contrast between treated and untreated phenylketonurics. The eleven year old boy is severely retarded, whereas his treated two and one-half year old sister is normal. (Reproduced from Pediatrics 2000; 105: 89-103).

There were several huge obstacles, however, that had to be overcome before the full benefits of the treatment could be realized. The methods used to fashion the special diets were not suitable for their widespread use. The method starting with a casein hydrolysate was tedious and the one starting with a mixture of pure amino acids was expensive. Luckily, this problem was solved when a variety of low-phenylalanine dietary products, such as Lofenalac, Cymogran, Ketonil, PhenylFree and Minafen became commercially available.

The other obstacle was far more serious. It was simply that at the time when this treatment was being developed no method was known that was capable of the very early detection of PKU. The scope and urgency of this problem can be appreciated from the evidence that the I.Q. of untreated patients was found to fall by about 4 points for each 4 weeks' delay in starting the diet. That meant that the diet started after a year's delay would lead to about a 50 point loss in I.Q. And such losses are irretrievable, gone forever. As noted by Pearl Buck, the Nobel Prize winning author, who was also the mother of a PKU child, PKU children who were treated late would never be what they could have been.

CHAPTER 6
DIAGNOSIS OF PKU

With the realization of how critical early treatment was, questions were raised about the adequacy of the urinary ferric chloride test for phenylpyruvic acid that Fölling had originally used to detect PKU. There were early indications that it was not up to the task. A serious limitation was that phenylpyruvic acid did not appear in the urine until about a month after birth, a lag that would limit the effectiveness of the diet. Even more worrisome was the finding that this acid did not begin to spill out into the urine until blood levels of phenylalanine rose to 10 to 15 times above normal levels. This far exceeded what was regarded as a safe level.

Both of these features of the excretion of phenylpyruvic acid shouted the message that the continued use of the ferric chloride test for the diagnosis of PKU would blunt the benefits of the dietary treatment. Clearly, one way around these limitations would be a screening test based on measurements of blood phenylalanine levels, which became abnormally high shortly after birth. But this path was blocked. There was no convenient quantitative test for the amino acid. There was nothing like the simple color test for phenylpyruvic acid.

In 1961 Robert Guthrie, an experienced microbiologist who had earned both an M.D. and a Ph.D. and who at the time was working in the Department of Pediatrics at the State University of New York at Buffalo, devised a clever semi-quantitative test for phenylalanine. Guthrie had become interested in mental retardation because one or his own children, John, was mentally retarded. Although John did not have PKU, being the father of a retarded child had stimulated Guthrie's interest in this area of medicine. A few years later, when he learned of the dietary treatment for PKU and of the necessity for frequent monitoring the treated patient's blood phenylalanine levels, he realized that a quick reliable test for phenylalanine was urgently needed. Whereas, Fölling, with his training in chemistry, had come up with a chemical solution to the problem of screening for PKU, Guthrie, with his training in microbiology, came up with a microbiological solution. As somebody once said, if the tool you are most familiar with is a hammer, you will tend to treat every problem as if it were a nail.

Relying on his training, Guthrie elaborated a test for phenylalanine that starts with a culture of bacteria whose growth is inhibited by thienylalanine, a synthetic analogue of phenylalanine known to behave like an "antimetabolite" against the amino acid in a variety of different biological systems. In the presence of the analogue, growth of the bacteria is rather specifically restored by the addition of phenylalanine; moreover, the extent of the growth is strictly proportional to the amount of phenylalanine added.

The test is carried out on a drop of blood obtained from babies, while they are still in the hospital, by a prick of each baby's heel. The individual drops are spotted on a piece of white absorbent paper, called filter paper, which looks and functions like blotting paper. The paper with its 1/8 to 1/4 inch diameter spots of dried blood is pressed down on a shallow dish containing the growth-inhibited bacterial culture mixed in with agar, a jello-like support substance. After overnight incubation of the dish at 37 degrees centigrade, the spots of blood from PKU babies are readily distinguished from those without PKU by the cloudy halo of bacterial growth around the spots from the PKU babies, due to the phenylalanine-mediated awakening of the dormant bacteria from their inhibitor-induced growth arrest. The amount of phenylalanine in the different spots is scored by a comparison of the size of the halo with that produced by standard amounts of the amino acid that were also spotted on the paper and carried through the procedure.

Because the pieces of filter paper can each accommodate blood spots from dozens of babies, and can be collected even from remote villages and mailed to medical centers around the world, the "bacterial inhibition assay", or, as it is more commonly known, the "Guthrie test", is tailor-made for mass screening of newborn babies for PKU. It has been widely adopted throughout the world. By 1967, the 48 contiguous states in the U.S. had PKU screening programs in place. In 1963, Massachusetts became the first one to pass a law requiring testing, a lead that was quickly followed by the other states. It has been estimated that more than 150 million infants now have been screened with the Guthrie test and over 10,000 have been detected with PKU and treated. In addition to this test, a fluorometric assay for phenylalanine has also been used frequently in newborn screening programs for PKU.

As of the year 2001, there was no uniform policy in the U.S. concerning the cost of the screening test or who would pay for it. A recent survey of the practices in different states showed that the cost varied from no fee (9 states) to $59 for each test. In the majority of states, the fee was between $20 and $40 for each test, with some states charging less for a second test or waiving the fee entirely.

Early screening for PKU with the Guthrie test made it possible for the first time to judge the full therapeutic potential of the dietary treatment of the disease. It was a triumph! The results showed that when started early enough- within the first weeks after birth- this therapy could save PKU infants from their mean genetic fate, could rescue them from essentially all of the devastating consequences of the disease, including the prevention of the relentless deterioration of their I.Q.s. The success of the treatment was a striking validation of the idea, first put forth by Francis Galton, an English scientist who was a cousin of Charles Darwin, that the sum of an organism's characteristics, known as its "phenotype", is determined by the interaction between "nurture and nature". The experience with PKU showed that the fearful consequences of a morbid nature, in this case, a mutant genotype, could be modified by a different nurture or environment, in this case, by how much phenylalanine is in the affected baby's diet.

In the early attempts to traverse the minefield leading to an effective therapy for PKU, not every mine could be avoided. There were casualties. Most of the problems arose because the goal of the treatment-to limit the intake of phenylalanine to the point where a normal blood level of the amino acid was reached- proved to be both unattainable and dangerous. Because the treatment bar was set too high, a new , artificial disease was created, "phenylalanine deficiency syndrome". Infants who were "overtreated" with the low phenylalanine diet suffered from a wide range of symptoms, including lethargy, vomiting, diarrhea, anemia, anorexia, edema, osteoporosis, growth retardation, low blood sugar and protein levels. Ironically, there was evidence that this kind of malnutrition during the first 6 months of life also resulted in permanent subnormal mentality.

Fortunately, before too long it was realized that by trying to achieve a normal blood level of phenylalanine, its level in the treated infants often fell well below the minimum needed for normal growth and development. With insufficient amounts of this amino acid, which is an essential part of most of the proteins in our body, just about every organ system was being slowly but surely starved to death.

The remedy was to simply lower the bar and liberalize the amount of phenylalanine consumed. Instead of trying to attain and maintain normal blood levels of the amino acid, it was found that keeping the level somewhere between twice and six-times normal (between 2 milligrams/deciliter, i.e.,120 micromoles/liter and 6 milligrams/deciliter, i.e., 360 micromoles/liter) gave good results. Even with this somewhat liberalized diet, blood levels of phenylalanine have to be monitored frequently, at least weekly during the first year or two of life. This means that during this period, even after the baby has been taken home it must remain at least loosely tethered to a clinic or hospital.

Sooner or later, every parent of a PKU child becomes fluent in the arcane language of "phenylalanine equivalents" of various natural foods. This is especially true as the child gets older and the diet can be liberalized by the addition of various foods to the low phenylalanine preparation. Extensive lists are available showing the phenylalanine content of most foods, usually calculated as the amount of any particular food that contains 15 milligrams of the amino acid. For example, this amount will be supplied in 11 tablespoons (tbsp.) of applesauce or 3 tbsp. of green beans or 1 tbsp. of oatmeal. For comparison, 1 medium egg contains about 300 milligrams of phenylalanine. The quantity of any food that can be added to the basic low phenylalanine diet is determined by its phenylalanine equivalents and the child's blood levels of phenylalanine. For those parents who do not want to be doing the arithmetic, various foods have also been roughly classified according to their phenylalanine contents:

High-phenylalanine foods (e.g., meat, fish, eggs, cheese, certain wheat products and some vegetables like beans). These are generally avoided.

Medium-phenylalanine foods (e.g., cream, milk, rice, corn, potato) which contain about 25 or 50 milligrams of phenylalanine for a usual serving size: a medium size potato , e.g., contains about 75 milligrams of phenylalanine). Within this category, exchanges can be made on a portion-for- portion basis. As far as milk is concerned, it is noteworthy that human milk contains much less phenylalanine (about 18 milligrams per ounce) than cow's milk (about 43 milligrams per ounce) providing still another reason in favor of breast feeding of infants.

Low-phenylalanine foods (e.g., most fruits and vegetables, and refined fat and carbohydrate). These usually contain less than 25 milligrams of phenylalanine for a typical portion. Prudent amounts of foods in this category can be eaten without too much restriction.

In addition to lists of the phenylalanine contents of various foods, a whole industry has sprung up that caters to the special dietary needs of PKU children. These companies produce a wide selection of low- protein (and therefore, low-phenylalanine) foods that add welcome variety to the diet of the older PKU child. The following list shows a typical day's menu for a PKU teenager on a liberalized diet.

Breakfast:

Cereal with a medical food preparation like Lofenalac

Lunch:

Salad or sandwich (lettuce, tomato, mayonnaise) with low protein bread

Potato chips or other chips

Fruit (banana, orange, grapes, cantaloupe)

Snacks:

Lemon pudding

Chips

Rice cakes with cream cheese

Baby carrots and dip

Eggplant dip

Dinner:

Pasta (low protein) with spaghetti sauce

Low protein bread

Rice or potatoes

vegetables and fruits

The special low-protein foods, however, are expensive, about 2 to 3 times more expensive than ordinary foods. Low protein pastas like spaghetti cost $3.75 for an 8.75 oz box, wheat starch bread costs $4.90 for a 17.6 oz loaf, low protein chocolate chip cookies cost $2.95 for a 6oz box and low protein porridge costs $8.45 for a 17.6 oz box.

Although the dietary treatment is a blessing, it imposes an appalling burden on the family of a PKU child. One of the problems with the treatment is that most of the commercially available low phenylalanine products like Lofenalac are notorious for their unpleasant taste. Fortunately, most babies generally adjust well to these products. The problem, however, can intensify as the babies get older and their tastes become more discriminating. A recent review of the general attitude toward these medical foods stated that "Most tasters judge these low phenylalanine products to be relatively unpalatable, with the PKU formula being rated as poor tasting and malodorous."

Within the last few years, there has been a noteworthy advance in the field of PKU formulas. NRM Buist and his colleagues at the Oregon Health Sciences University, Portland, have succeeded in formulating a new amino acid mixture for treating PKU patients that appears to be more palatable than those that were previously available and yet manages to maintain the plasma pattern of amino acids (except for phenylalanine) well within the normal range. They achieved this goal by eliminating some of the non-essential amino acids such as glycine, glutamic acid and aspartic acid and decreasing the amounts of the sulfur- containing amino acids, methionine and cystine from the mixture. The acceptability of the new product was found to be "markedly improved." As of 1994, this product was available from the clinic in Oregon.

Patience and a thorough understanding of what is at stake in making the dietary treatment work are the essential ingredients for successfully avoiding the Scylla of providing too much phenylalanine and the Charybdis of providing too little. Feeding problems are exacerbated when the child is ill and refuses to eat the low phenylalanine preparation. The whole experience of tending a PKU infant can leave the parents very stressed . Proper counselling is needed to make sure that the price paid for treating the child is not a divorce. On top of all that, the low phenylalanine diet is expensive, costing between $5000 and $10000 a year. In most states, at least part of this cost is covered by insurance.

The early experience with the dietary treatment indicated that the diet could be discontinued after the age of about 4 years with little or no further deterioration in mental development. This was certainly welcome news since it meant that these treated kids could go off to school without the stigma and the fuss associated with having to eat special food. In the world of the PKU child, where there was very little to be thankful for, the recommendation that the diet could be stopped before school started was something quite precious for the whole family. It seemed to signal the end of a long ordeal.

But this hopeful signal proved to be premature. The first hints that all was not well with the group of PKU kids who had been treated but were now off the phenylalanine restricted diet came from the school teachers. They reported that many of these children were disruptive in class and had learning problems. These worrisome anecdotal reports were soon followed up with the results of controlled clinical studies which told the same story. Most, but not all, PKU children who had been successfully treated with the diet since early infancy showed significant deterioration in their performance after they had been taken off the diet sometime between ages 5 and 6 years. The changes included modest decreases in their IQ scores (5 to 30 points), as well as deviant EEG findings, deficits in their higher cognitive reasoning abilities, impaired reaction times and assorted behavioral problems.

Although there seemed to be clearcut benefits to staying on the diet at least until the age of 8 compared to going off the diet before the age of 6, there was a surprising lack of uniform guidelines telling parents what to do beyond that period. Given that state of uncertainty, and the unpleasantness of the treatment, it is not surprising that some parents took it upon themselves to terminate it whenever they decided that they'd had enough. When, in the late 1970's, the depressing recommendation came out that the diet, or a somewhat liberalized version of it, should probably be continued for life, patients who had discontinued the diet found it very difficult to reinstate it. The rewards for resuming it, however, were great. As will be discussed later, it was particularly important for PKU women to stay on the diet through their childbearing years. If they did not do this, the mother's elevated blood phenylalanine levels would damage their unborn babies and lead to a condition called "maternal PKU".

Later related studies provided strong support for the notion that some of the adverse effects of hyperphenylalaninemia are reversible. These studies showed that it was possible to acutely induce a poor performance on tests that measure higher integrative function (such as attention span, visual or auditory memory, ability to follow instructions) by feeding PKU patients a load of phenylalanine. Performance improved again when blood phenylalanine levels returned to normal, an indication that these acute bad effects of high phenylalanine were reversible.

These last results were coherent with those from the early experience with the dietary treatment for PKU. It may be recalled that with children who were placed on a phenylalanine restricted diet after a delay of years or even months, there were some beneficial effects but these did not include preventing a serious drop in IQ. These results showed that there are 2 types of symptom in PKU: those that are reversible by phenylalanine restriction such as the deficit in pigment formation and those that are irreversible but largely preventable, such as the mental deterioration. There are indications that different mechanisms are responsible for the reversible and the irreversible symptoms.

During the decade following the introduction of the dietary treatment for PKU, the reports on the outcome continued to be rosy. They seemed to show that, when started early enough- within the first weeks after birth- this therapy could rescue the PKU patient from all of the devastating consequences of the disease.

This first flush of success with the low phenylalanine diet was followed by indications that despite the clear benefits of early treatment in blunting the full clinical impact of the disease, the treated PKU patient is not entirely normal. In addition to small deficits in IQ, the majority of these kids have a higher than normal frequency of learning difficulties (e.g., slowed response times, delay in acquisition of language and impaired problem-solving ability), as well as behavioral problems such as hyperactivity, anxiety, and poor concentration.

This list of the way even well treated PKU patients are seen to deviate from normal agrees fairly well with the way these patients view themselves. Compared to normals, a group of German PKU patients reported that they were significantly less satisfied with life, less socially oriented, less motivated towards achievement and success, less open, less extraverted, less emotional and that they had a lower frustration tolerance. On their own admission, 59% of this group of patients stated that they were unable to manage the diet without help from their mothers. Just how crucial the mother's role is in managing the diet can be seen from the finding that low socio-economic status and low IQ of the mothers were significantly correlated with higher average serum phenylalanine levels of the patients (probably a reflection of poorer dietary control). Furthermore, and not surprisingly, patients with poor dietary control had more social and emotional problems. Also not surprisingly, the majority of the mothers in this German study characterized the upbringing of their PKU children as overprotective and restrictive.

Another study of parents of PKU children gives us a glimpse of the full impact of dealing with this illness and its treatment. Two themes tower above all others. One is the initial shock on learning about the diagnosis and the other is the chronic problem of dealing with the special diet.

One of the mothers described her experience as follows "....I had been home for 10 days when the hospital rang up to say they needed more blood and urine from N., something was wrong.....Then the doctor said we should come in right away but it would take a week, so N. would have to stay in the hospital. That was like a whip-lash for me.... I was stunned and couldn't do anything because he had been so blunt about her having to stay in for a week."

Another mother gave this account. 'Well, the whole of the first 6 months were certainly overshadowed by the way in which we were informed about the illness itself. When we heard that the test was positive we were quite at sea for 10 days-really in despair-because the first explanation we had heard was that the idiot test was positive, and perhaps something could be done...." (One can only hope that the emotionally-charged term "idiot" is no longer used in this context).

There are some indications that the narrative descriptions of the experience of these two German mothers may not be representative, that they may paint too gloomy a picture of the impact on the family of having to cope with a PKU patient. Indeed, in this context, it should be emphasized that a discussion of the imperfections in the treatment for PKU should not be allowed to obscure the consensus view that "most early treated children fall within the broad normal range of general ability and attend ordinary school".

I recently spent a few hours chatting with one of these success stories, an attractive, bright 25 year old women with classic PKU who had been treated with a phenylalanine-restricted diet since early infancy. I met with her and her parents at their house in the Maryland suburbs. There was nothing remarkable about the patient except perhaps her apparent normalcy. She was about to complete her 2-year graduate training period for a degree in Pharmacy. She already had a job lined up as a pharmacist in a local supermarket. I was surprised to learn that she was no longer on a strict diet, although she still limited her intake of high phenylalanine foods like eggs, meat and fish. Interestingly, she had not noticed any adverse effects on switching to a more liberal diet. She knew patients of her age whose parents did not allow them to go off the strict diet. One of them, who was attending an out-of-state college, was still being sent all of his specially prepared food by his mother. I was also surprised that the young lady whom I had visited had not had her blood phenylalanine level checked for a few years. As the mother mentioned, this was obviously far different from the time when she was an infant and a nurse from the State would come to their house once a week to collect blood, a practice that was continued for the first 2 years of the baby's life. By contrast with the happy outcome for this young woman, before the development of the dietary treatment the vast majority of PKU patients were destined to spend their lives not in an ordinary school but rather in a mental institution.

Most parents of PKU children try to explain to their kids the nature of their illness. Rather than telling them that the cause of their condition is that they lack an essential enzyme in their liver, the parents become quite creative in inventing non-technical explanations. I heard the following story about a parent who told her five year old girl that "an enzyme is like the juices in her mouth that helped make crackers soft and dissolve." A couple of months later, the mom overheard her telling her violin teacher, "I have PKU, which means that I have a juice missing from my liver". The institution of the low phenylalanine dietary treatment for PKU impinged on many aspects of life of the families of the PKU patients. One of the aspects that was affected was the decision about whether breast-feeding was allowed.

In fact, the attitude toward breast-feeding in the management of infants with PKU has had a checkered history. Shortly after the low phenylalanine diet was introduced as a treatment for the disease, experts warned that breast feeding was not compatible with attempts to control the patient's hyperphenylalaninemia with a low phenylalanine diet. More recently, however, the pendulum has swung completely. Currently, breast feeding is not only allowed, it is actually encouraged. The change in attitude was propelled by the realization that human breast milk is one of Nature's most nearly perfect foods for the PKU infant. That is because, as mentioned previously, human breast milk contains less than one-fourth the amount of phenylalanine compared to cow's milk. Breast feeding does involve some additional attention in order to keep track of the total amount of phenylalanine consumed by the infant. One way to do this is to weigh the baby before and after each breast feeding and from the volume of milk consumed and the known phenylalanine content of breast milk to calculate the amount of phenylalanine that had been consumed. Based on this value, an amount of phenylalanine-free formula is offered to the infant to provide the amount of energy and protein previously shown to be needed for what is judged to be ideal growth.

Despite the great success of the newborn screening program for PKU and of the dietary treatment for the disease, the program has drawn fire from some bioethicists. The target of the attack is that since the 1960's, newborn screening for PKU has been mandated by law in most states. This mandate appeared to breech the long-standing rule that governs, at least implicitly, the practice of health care in America, the rule of informed consent. The mandate appeared to take away from parents the right to decide whether or not their children would be tested for PKU and, if necessary, treated for the disease. Parents were brought back into the picture in 1994 by the recommendation by the Institute of Medicine that "no genetic test should be done without the consent of the persons being tested, or in the case of newborns, the consent of the parents."

This recommendation was on a collision course with the law mandating PKU screening. It created a potential nasty moral and legal dilemma. What would happen if parents refused to give consent to have their children tested and treated for PKU? A group of bioethicists at Johns Hopkins University has recommended that in this situation legal steps should be taken to compel parents to treat and, if necessary, to remove the infant from parental control for treatment through statutory provisions covering child neglect. I am not aware of any case in the US where this has actually happened.

There are those who have argued that it would not be unprecedented if a parent were to decide against allowing neonatal screening and treatment for PKU. Two main arguments have been offered to support such a decision. First, it has been pointed out that there is a serious down side to the Guthrie test for PKU, that it generates too many false positives, i.e., infants who tested positive for PKU and who were therefore started on the low phenylalanine diet only to find out later that they really did not have the disease. Not only were some children needlessly harmed by the treatment (probably mainly from "overtreatment" when their blood phenylalanine levels were allowed to fall too low), but their parents were subjected to unnecessary worry and stress. A solution to this potential problem would be a requirement for a confirmatory positive test before the low phenylalanine diet is initiated. As already discussed, the specific problem of overtreatment largely disappeared when a higher level of blood phenylalanine became acceptable as the goal of the treatment.

A second argument that has been marshalled against mandatory screening for PKU is that a parent's decision not to allow screening would only result in the exposure of their child to about a 1:10,000 risk of serious harm. Society routinely allows parents to make decisions that expose their children to comparable or even higher risks. This happens every time a parent allows their children to ride a bicycle or engage in gymnastics.

As this brief discussion indicates, there are examples of parents making decisions that subject their children to a risk of serious harm. In most of these examples, however, there is a clear downside to the procedure or the event that the parents reject. When parents do not prohibit their children from riding a bike, even though they know that bike riding can be dangerous and even deadly, they have decided that depriving their child of the pleasure of riding a bike outweighs the risks attached to this activity.

But what is the downside to neonatal PKU testing? Historically, the principle risk that has been cited is the frequency of false positives. Indeed, in the early days of the screening program the number of false positives vastly exceeded the number of true positives. As mentioned above, the hazards of needlessly treating these children and possibly harming them could be minimized by the expedient of repeating the Guthrie test or by the use of different test such as a chemical (fluorometric) quantitative assay for phenylalanine.

Whereas repeat testing minimized the problem of false positives, it did not eliminate it. An analysis of the factors that influence the occurrence of false positives identified testing that was carried out too early as a major factor. That, in turn, was partially due to the increasing trend toward early discharge of newborns from nurseries. To avoid the inconvenience of having a parent bring the infant back to the hospital for testing and to facilitate the earliest possible initiation of the special diet if the infant tested positive, there was an understandable push to carry out the test before the infant left the hospital. This largely societal pressure created a medical problem. It not only set the stage for the screening procedure to pick up more false positives, but also more false negatives, i.e., PKU cases being missed. Adoption of this practice of very early screening, therefore, imposed a double handicap on the program.

The reason for this unfortunate situation is that blood phenylalanine levels in authentic PKU patients at birth are far below, often only one tenth, their ultimate peak; these levels continue to increase through the first 14 days of life. It has been estimated that there would be a 16-fold improvement in the reliability of the screening for PKU when the blood sample is drawn on the third day of life (48 to 72 hours) compared to the second day of life (24 to 48 hours). As a result, the proportion of babies who are missed would be expected to increase from 0.15% to 2.4% when the test was carried out on the second day rather than the third day. Early screening, coupled with this pattern of delayed development of blood phenylalanine levels, can explain why screening carried out too early generates more false negatives. The chances of missing the diagnosis is greatest in infants screened at the youngest ages.

Ironically, the chances of picking up false positives are also increased when the testing is carried out too early. The reason for this confounding result is that non-PKU infants who happen to have either low birth weights, or who are premature, tend to have transient defects in tyrosine metabolism. As a result, these infants tend to have elevated blood levels of tyrosine, often accompanied by high levels of phenylalanine. This transient increase in blood levels of both of these amino acids in this group of infants opens the door for false positives to enter the picture. Because they lack the ability to convert phenylalanine to tyrosine, authentic PKU patients, even those who have low birth weights or are premature, have low, rather than high, blood levels of tyrosine.

The recent exciting development of sensitive sophisticated new methods for accurately determining both phenylalanine and tyrosine in dried blood samples has largely eliminated the problems inherent in early screening. One of these new methods is based on the determination of the amino acids by a procedure called tandem mass spectrometry or MS/MS. The procedure is sufficiently accurate and sensitive that it has detected PKU in samples collected from newborns even younger than 24 hours of age. Even more impressive, in a recently reported study the MS/MS procedure has been shown to have eliminated 90 out of 91 false positives that had been picked up previously by the fluorometric screening test.

The MS/MS procedure is still in its infancy but it currently is used for all the PKU screening in Pennsylvania and the District of Columbia, as well as a significant fraction of the screening in Massachusetts and North Carolina. With its compelling advantages, the odds are that the MS/MS procedure will eventually clear the field of all other PKU screening tests. And, when the problem of false positives is essentially eliminated by this procedure, it is difficult to imagine why any parent would not consent to having their child screened for PKU. Hopefully, at that time the whole troubling issue of informed parental consent for testing for PKU will fade away.

The early detection of PKU and the subsequent development and institution of the low phenylalanine diet to prevent the mental retardation associated with PKU, have been a huge success. Indeed, as stated by Frank Lyman, M.D. in 1963, they represent one of the great advances of medicine. But the method used for the detection of the disease- a blood test carried out within a few days of the infant's birth that measures blood phenylalanine levels- is a rather blunt instrument. By itself, it cannot be used optimize the dietary treatment to the needs of the individual patient.

One of the reasons for the need for a more powerful method is the complexity of the clinical picture of PKU, a complexity that derives in part from the unexpectedly large number of mutations in the gene for phenylalanine hydroxylase, currently about 400 and still counting. Different mutations in the hydoxylase molecule can lead to different symptoms in the patient.

In an attempt to cope with this complexity, various classification schemes have been proposed to try to relate blood phenylalanine levels to severity of the disease and to clinical outcome. In one scheme, serum phenylalanine levels, measured before the baby has been put on a phenylalanine-restricted diet, above 1200 µM (20 mg/dl) has been defined as classical PKU, levels between 600 and 1200 µM has been called mild PKU and levels below 600µM has been defined as mild hyperphenylalaninemia or non-PKU (which means elevated serum phenylalanine levels without the urinary excretion of phenylketones such as phenylpyruvic acid). Because blood or serum levels of phenylalanine in this disease represent a continuum, classification schemes such as this one, although they are useful, have been recognized as being too dogmatic.

Genotype-phenotype correlations

The Holy Grail in attempts to classify the great variety of hyperphenylalaninemias would appear to be one based on genotype-phenotype correlations. The goal here is to be able on a drop of blood, taken while the infant is still in the hospital, to identify the kind of mutation in the hydroxylase gene that is responsible for the infant's genotype and to correlate the genotype with the severity of symptoms in the patient (i.e., the patient's clinical phenotype). Ultimately, the goal of this effort is to be able to customize and optimize the dietary treatment to the needs of the individual patient.

The large number of mutations in the hydroxylase gene, as well as the phenomenon of "compound heterozygosity", which has been discussed previously, complicate any attempt to work out this kind of classification. It may be recalled that normal individuals have two wild-type hydroxylase alleles, each located on two homologous chromosomes, and PKU patients have an homozygous genotype in which each homologous chromosome harbors the same mutant alleles. In contrast, in compound heterozygotes the two chromosomes have different mutant hydroxylase alleles. In this last group, the clinical phenotype often will be determined by how these two mutant alleles interact.

Despite a considerable effort to work out generally applicable diagnostic procedures based on genotype-phenotype correlations, the present consensus view appears to be that this goal has not yet been reached. Nonetheless, some useful conclusions have come out of such studies. The approach that has been used has tried to correlate the severity of a mutation in the hydroxylase gene with the severity of the symptoms in the patient being studied. The severity of the mutation was ascertained by a technique known as "*in vitro* expression analysis". In this technique, a clone of mutant hydroxylase, created by mutagenesis of the hydroxylase DNA, is inserted into mammalian cells in tissue culture. After incubating these cultured cells in vitro, the hydroxylase activity that has been expressed in the cells is measured. With this procedure, the residual hydroxylase activity of a large number of naturally-occurring mutant forms of the hydroxylase have been determined and correlated with the severity of the symptoms in those patients known to harbor these mutated hydroxylase genes.

Mutations were classified as "severe" if expression analysis did not detect any hydroxylase activity, as "moderate" if between 1 and 10% of normal ("wild-type") activity was detected, and "mild" if more than 10% of normal activity was detected. The results showed that mutations such as R408W - in which the amino acid arginine (R), normally in position 408 in the hydroxylase protein is replaced by another amino acid, tryptophan (W)-is classified as "severe", i.e., in expression analysis it displayed essentially no (less than 1% of that seen with the wild-type) phenylalanine hydroxylase activity; in patients this mutation was shown to be associated with a severe form of PKU. By the same kind of analysis, mutation R408Q, in which the amino acid arginine (R) in position 408 in the protein is replaced by another amino acid glutamine (Q), has been shown to be a "mild" mutation and has been shown to be associated with mild PKU.

In cases like these two , where the results are clearcut, this kind of analysis can provide useful guidelines for the treatment of these patients. The patient with the severe mutation (R408W) usually will require stringent restriction of dietary intake of phenylalanine, whereas the patient with the mild mutation (R408Q) usually can be treated with a more relaxed restriction of phenylalanine intake.

Unfortunately, many of these attempts to determine the genotype-phenotype relationship have yielded discordant results. One of the more troubling findings is that the same mutant genotype is not always associated with the same phenotype in different patients. For example, based on results of the genotype-phenotype analysis carried out at 2 different medical centers, mutation R261Q has been assigned to 2 different phenotypes, either classic PKU or moderate PKU. The explanation for results such as these is not known with certainty. Although discordant results may limit the power of this kind of analysis, this approach is still useful for predicting a patient's phenotype and for customizing the treatment to maximize the benefits to the patient.

CHAPTER 7
MATERNAL PKU

Before the institution of routine newborn screening and early treatment for PKU, almost all women with this disease who reached childbearing age were severely retarded; not surprisingly, few of them got married or had children. Parenthetically and sadly, the fathers of many of the children born to these women were male relatives. The children born under these circumstances provided an ominous glimpse of future developments in this field. For although the great majority of the children of PKU women are not themselves phenylketonuric (most are heterozygotes), they are nonetheless mentally retarded and damaged in other ways-microcephaly, congenital heart defects, urogenital abnormalities, deformed facial features, low birth weight and slow postnatal growth, with mental retardation and microcephaly being the most common, affecting between 70 and 90% of the offspring. Most PKU heterozygotes are born to families where both parents are themselves PKU heterozygotes. On average, one half of the babies in such families will be PKU heterozygotes. These PKU carriers are therefore exposed *in utero* to only the modestly (about 50% above normal) elevated blood phenylalanine levels of the heterozygous mother. This is in sharp contrast to the situation faced by babies born to mothers with untreated classic PKU who are exposed to blood phenylalanine levels that are 15-20 times higher than normal. There is no doubt that it is the exposure to such high phenylalanine levels *in utero* that damages the offspring of these PKU mothers. The frequency of the abnormalities seen in these babies correlates with the mother's blood phenylalanine levels. For example, microcephaly, which has an incidence in the normal population of 2.7%, at phenylalanine levels of 3-10 mg/dL has an incidence of 24%, at phenylalanine levels of 11-15% an incidence of 35% and at phenylalanine levels greater than 20 mg/dL, an incidence of 73%.

Results published by Levy and Waisbren led these investigators to conclude that there is a threshold of phenylalanine levels of 10mg/100ml (0.6mM) and that only levels higher than this are damaging. This interpretation of the data has been challenged by Kirkman and Hicks at the University of North Carolina and by Buist and his colleagues at the Oregon Health Sciences University. As will be discussed in a later chapter, Isobel Smith in London has also weighed in on this issue, presenting data against the "threshold theory". Although the issue remains unresolved, the fact that there are opposing views on this aspect of the disease should serve as a warning about the safety to the fetus of even modest increases in the levels of phenylalanine.

The condition is called "maternal PKU" or "maternal phenylketonuria embryopathy". It was first recognized by Dent in 1957 and described in greater detail a few years later by Mabry and his colleagues, at the University of Kentucky School of Medicine. The facial features of babies with maternal PKU are reminiscent of those suffering from fetal alcohol syndrome. Studies of this condition supported the conclusion that phenylalanine is teratogenic (Greek, *teras* , marvel or monster). The offspring of PKU men appear to be perfectly normal, a finding that is entirely consistent with the idea that the disease is caused by the fetus developing in a high phenylalanine environment of the PKU mother.

The babies are damaged *in utero* by the PKU mother's high levels of phenylalanine at a time when major organ systems like the heart are being formed. A feature of normal phenylalanine metabolism that works against the developing fetus in maternal PKU is the usual gradient of amino acids, including phenylalanine, that favors the fetus by a ratio of about 1.5. This means that when the mother's blood phenylalanine levels are 400µM, the level in the fetus will be about 600µM. Whereas this feature is valuable to the fetus when it is developing in a normal *in utero* environment, since it provides the fetus with more amino acids during periods of rapid growth and rapid protein synthesis, it exacerbates the potential damage that can be caused by the PKU mother's excess phenylalanine. The mechanism by which excess phenylalanine damages the fetus is, in all likelihood, similar to the way it damages the developing brain in PKU infants. Just as is the case in that condition, there is strong evidence that it is phenylalanine, itself, rather than one of its metabolites such as phenylpyruvic acid that damages the fetus in maternal PKU. Excess phenylalanine is probably harmful to the fetus because it competes with the other large neutral amino acids for placental transport as well as for uptake by fetal tissues. As a result of this competition, the fetus is starved for these essential growth ingredients. There may also be a small decrease in the amount of tyrosine supplied to the fetus because of the PKU mother's inability to convert phenylalanine to tyrosine. Below, we will examine further the possibility that a relative deficiency of tyrosine contributes to fetal damage in this condition. Whatever the cause, the damage to the fetus is irreversible. Consequently, maternal PKU babies cannot be treated successfully with a low-phenylalanine diet.

Because of the comparatively small numbers of affected children involved, the problem of maternal PKU was for years a minor one. It proved to be, however, a vicious sleeping giant, which was aroused by the successful treatment for PKU. The dietary therapy enabled PKU women to lead essentially normal lives with normal marital and childbearing expectations. But, during that period when it was being recommended that the dietary treatment could be terminated after 6 or 8 years, most of the children of the women who had gone off the diet were born to women with substantially elevated blood phenylalanine levels. As a result, many of these babies were damaged. The great success of treating PKU had converted a relatively minor medical problem into a potentially major one.

Prevention of maternal PKU is, in theory, easy; in practice, it is daunting. The rational way to prevent it is for PKU women to return to the low phenylalanine diet *before* they become pregnant. This timing is crucial because high phenylalanine levels begin to cause damage to the fetus during the first trimester of pregnancy. Optimal fetal growth occurs only in infants whose mothers had phenylalanine levels close to the normal range at conception. Dietary treatment started after conception, say, when the woman first realizes that she is pregnant, is dramatically less effective than treatment started before conception; microcephaly, for example., can still occur. It is regarded as a medical emergency when a PKU woman first goes to her doctor when she is already pregnant. In some medical centers, if conception occurred when the woman's blood phenylalanine levels were persistently higher than 800µM (14 fold higher than normal) she is considered to have a strong case for termination of the pregnancy. On the other hand, it is generally accepted that if the PKU mother can get her blood phenylalanine levels down to a relatively safe range-less than 400µM- within the first eight weeks of pregnancy the chances of her having a healthy baby are still quite good.

In view of the problems associated with resuming the diet once it has been terminated, the occurrence of maternal PKU provides a powerful added argument for PKU women to continue the dietary treatment through their childbearing years. Indeed, the difficulties of resuming the diet are so great that it is recommended that women who have discontinued the low phenylalanine diet resume the diet at least 3 months *before* a planned pregnancy so that they can establish nutritional control and become accustomed once again to the drastic changes in their eating habits.

The inclination of women to resume the low phenylalanine diet once they have given it up will depend, among other things, on how inoffensive the therapeutic diet is. Fortunately, several products have been developed recently that address the problem of the poor palatability of the earlier low phenylalanine diet preparations. One of these is the formulation developed by N.R.M. Buist and his colleagues at the Oregon Health Sciences University has previously been discussed in connection with the dietary treatment of PKU. Another product achieves the desired goal of masking the offensive taste of the amino acid mixture by gelatin encapsulation of the mixture. With both of these preparations compliance has been found to be good as has been the fetal outcomes.

While there is little doubt that a properly supervised low phenylalanine diet can be effective in preventing fetal damage in maternal PKU, there are several issues related to this treatment that need to be examined. One of these is the possibility that, in addition to restriction of their phenylalanine intake, L-tyrosine supplementation might be beneficial in the treatment of PKU pregnancies in those women whose plasma tyrosine concentration is below the concentrations considered to be appropriate for optimal fetal development (45μM). This possibility has been raised by Harvey Levy and his colleagues on the basis of their observation that the mean plasma tyrosine concentration of a small group of PKU women was only 35μM, less than the 45μM considered to be optimal for fetal development. On this point it is worth mentioning that the tyrosine content of the more palatable phenylalanine-free amino acid mixture developed by Buist and his coworkers is twice as high as that in PhenylFree, one of the older amino acid preparations that has frequently been used in the treatment of PKU children. This newer mixture, therefore, has already been supplemented with some extra tyrosine. Two additional points should be noted about tyrosine supplementation. First, there is no evidence that the addition of extra tyrosine to the diet of PKU mothers would be beneficial to the fetus. Second, tyrosine supplementation must be monitored with care because too much tyrosine can, by interfering with the cellular uptake of some of the other amino acids, lead to serious problems for the fetus.

With respect to the question of how to optimize the dietary treatment for maternal PKU, there is some evidence suggesting that improved intakes of vitamin B12 and folate may decrease the risk of congenital heart disease even in the presence of elevated blood levels of phenylalanine.

During the 3 month adjustment period, referred to above, and the 9 months of pregnancy the cost of the special diet can add up to $4500. As with the treatment of PKU infants, pregnant PKU women on the diet must have their blood phenylalanine levels checked very frequently, at least twice a week The goal is to keep the fetal phenylalanine levels below 500 µM, which means that the mother's level must be kept below 300µM to 400µM. To achieve this goal, the pregnant woman's intake of natural protein must be decreased to 6g/day, or less, and at least 70 g of supplemental amino acids per day must be eaten to cover nutritional requirements. From around 20 to 22 weeks into the pregnancy, phenylalanine tolerance increases to as high as 30 g of protein per day, presumably because of the development of significant levels of hepatic phenylalanine hydroxylase in the heterozygous fetus that help in disposing of some of the phenylalanine.

Just how formidable a problem it is for some PKU women to resume the low phenylalanine diet after they have given it up can be seen from a sad case that was recently reported in the medical literature. This young woman first went to see her physician in the eighth week of her pregnancy when her blood phenylalanine levels were already 15-fold above normal. In addition to this hazard to her fetus, she was also smoking about 15 cigarettes a day. Amazingly enough, she was able to stop smoking for the duration of her pregnancy but was unable to eliminate meat from her diet. Consequently, her blood phenylalanine levels remained grossly elevated throughout her pregnancy. Not surprisingly, her baby showed many of the signs of maternal PKU, including growth failure, developmental delay, and slow acquisition of language. He is mentally subnormal and attends a school for the intellectually impaired. Sadly, two subsequent pregnancies, one a boy and the other a girl, had very similar outcomes, almost certainly because, again, the mother was unable to give up eating meat The mother's explanation for her persistence in continuing to become pregnant was her desire to have a daughter. Having achieved that goal, she agreed to be sterilized, bringing an end to this sad story. The bottom line? PKU women should give serious consideration to staying on the low phenylalanine diet through their child-bearing years.

With the development of several more palatable artificial low phenylalanine amino acid mixtures and the probability that these products will make it easier for PKU women who are planning to become pregnant to resume the diet, one of the major obstacles to the successful treatment of maternal PKU has been removed. There are, however, other serious problems remaining. One of the most formidable challenges in the field of maternal PKU is a logistical one rather than a purely scientific one. It is the task of identifying as many as possible of those women in the reproductive age group with hyperphenylalaninemia and advising them that before they become pregnant their plasma phenylalanine levels should be brought down to the normal range or, at least down below 400 uM.

The group of women at greatest risk are those who were born before mandatory PKU screening was required in their states or localities. Since newborn screening for PKU was begun in the United States between 1962 and 1985, many women who were born before 1962 and even some who were born as late as the mid eighties were probably not screened and may therefore have blood levels of phenylalanine that are well above normal. Should they become pregnant, their phenylalanine levels may be high enough to cause fetal damage. Even among those women who were screened at birth, a surprisingly large number of them were "lost to follow-up". These women are also likely to become pregnant without sufficient attention being paid by their health care providers to their phenylalanine blood levels.

The number of women whose children are at risk because of maternal PKU is staggering. In 1994 in the United States, 1,341,602 live births were to women older than 30 years. It has been estimated that as many as 40% of women currently (in 1999) bearing children in North America and

Europe were not screened for PKU as neonates. If one uses the same figure for the whole decade prior to 1994 and assumes that one-half of the births occurring during this period were females, one comes up with a figure of about 6.5 million (10X1,341,602/2= 6.5 million) women who were never screened for PKU and who are still in their child-bearing years. If the incidence of PKU in this group of women is 1:12,000 births, they would give birth to between 500 and 600 babies with undiagnosed maternal PKU. Fortunately, with this aspect of the disease, time is on our side, because with each passing year, more and more women of reproductive age will have been screened for PKU as neonates, thereby steadily decreasing the number of women whose PKU was undiagnosed.

How can these women be identified before they place their unborn children at risk? Hanley and his colleagues have concluded that universal PKU screening, which would be the most certain way of identifying them, would probably not be cost-effective. Instead, some kind of selective screening or case finding might be a better alternative. One might think that targeting women who are mentally retarded would be a good place to start, since without neonatal screening and treatment of those who tested positive for PKU a significant number of them would be retarded. It turns out that this criterion would probably not be as useful as might be expected because, by some estimates, as many as 20% of subjects with untreated classic PKU may have intelligence within or near the normal range. For those with mild hyperphenylalaninemia, in contrast to those with classic PKU, the number with notable mental handicap may only be about 50%. An urgent question is whether infants born to mothers who have only mild hyperphenylalaninemia during their pregnancies are at risk? Although a definitive answer to this question is not yet available, preliminary data indicate that even mild maternal hyperphenylalaninemia (plasma phenylalanine levels between 365 and 600 µM) may not always be benign. For example, it was reported by Isobel Smith in 1987 that for each 200uM increment in the pregnant woman's blood phenylalanine levels there is a 0.5 centimeter deficit in the infant's head size. This would mean that even a modest degree of hyperphenylalaninemia, leading to an increase in blood phenylalanine levels from the average normal level of 60µM to 260 µM could lead to a modest degree of microcephaly. The indications that such mildly elevated plasma phenylalanine levels, which are not associated with mental retardation in the mother, may nonetheless damage the fetus, might appear to be paradoxical. These findings suggest that the fetus suffers more damage than does the mother, an indication that the rapidly growing fetal tissue is probably more vulnerable to phenylalanine toxicity. Related to this point, it has been noted that damage occurs to fetuses exposed to other toxins, such as cocaine, alcohol, and polychlorinated biphenyls, whereas postnatal exposure to these compounds does not produce the same damage.

If one accepts the conclusion, mentioned above, that universal prenatal maternal PKU screening would not be cost-effective, what are the alternatives? Hanley and his associates have outlined a sensible game plan for dealing with this problem. They have recommended that there should be mandatory screening for women considered to be at the highest risk. These include women with a family history of PKU, women who have already given birth to a child with microcephaly or mental retardation and women who are known to be retarded or of borderline intelligence.

Serious consideration should be given to screening all women who were born before neonatal screening was introduced in their country or locality, as well as women whose offspring had congenital heart disease. Ideally, these women should be screened before conception.

Finally, there are women who probably do not need to be screened. This group would include those who were born after neonatal screening had been started in their jurisdiction. Women who have previously given birth to an unaffected child would also fall outside the group needing to be screened, as would women whose racial or ethnic background makes PKU very unlikely (Afro-Americans, Japanese, Finns).

In order for selective screening to be effective, it is essential that physicians in obstetric practice and midwives be alerted to the need to identify patients who are at high risk.

A few years ago, Professor Charles Scriver of McGill University in Montreal assessed the situation with respect to maternal PKU as follows: "Maternal hyperphenylalaninemia continues to be a major challenge despite more than 30 years' awareness that it would become one in the history of PKU. It is a problem that must be resolved; otherwise, all achievements in the prevention of mental retardation in PKU will have been gained at the cost of a terrible Faustian bargain. "

CHAPTER 8
VARIANTS OF PKU

I have already discussed how the complex phenylalanine hydroxylating system was successfully dissected into its individual components, phenylalanine hydroxylase, the reductase and the non-protein coenzyme tetrahydrobiopterin. On completing this work, it was obvious to me that there might exist different forms of PKU-variants- each caused by the lack of one of the three essential components of the system. In addition, it could be predicted that because the reductase and the coenzyme were also known to be essential for the activity of two enzymes, tyrosine hydroxylase and tryptophan hydroxylase, that catalyze the synthesis of key neurotransmitters in the brain, noradrenaline, dopamine and serotonin, that patients lacking either of the latter components of the hydroxylating system would probably be much more seriously ill than are PKU patients. Such patients, after all, unlike those with PKU, would suffer from at least three distinct metabolic lesions.

I made this prediction in 1967 but almost a decade would pass before there were any reports that such variants actually do exist. This period of relative calm ended in 1974-1975 with a couple of bolts of lightning in the form of two publications that appeared within a span of only six months, one from Isobel Smith and her coworkers in England and the other from Klaus Bartholomé in Germany. These articles described five PKU patients suffering from a progressive neurological illness which did not respond to early treatment with a low phenylalanine diet. This alone would have indicated that these children did not have the classic form of the disease. This suspicion was confirmed when hepatic phenylalanine hydroxylase levels measured in a few of the patients turned out to be within the normal range, proving that they had a new, and what proved to be a deadly, form of PKU. In one of the studies involving 3 children, all of them died between the ages of 22 months and 7 years, from what at first may have seemed like a fairly benign symptom-they had swallowing difficulties. This problem, however, had a deadly consequence, namely, that they often aspirated their food, an event that frequently led to bronchopneumonia. To dramatically distinguish this illness from classical PKU, this new form was called "lethal PKU" or "malignant PKU". There was speculation that the underlying metabolic problem in these children was some defect in the metabolism of the coenzyme, tetrahydrobiopterin. These reports were alarming because at that time no treatment was available for this deadly disease.

Hyperphenylalaninemia caused by dihydropteridine reductase deficiency

Sometime early in 1975, I got a telephone call from Dr. Neil "Tony" Holzman, in the Department of Pediatrics, Johns Hopkins University School of Medicine in Baltimore, just 45 miles away from NIH where I was working. I had met Tony at scientific meetings on PKU; he was therefore familiar with my work on the phenylalanine hydroxylating system.

Tony was calling about an atypical PKU patient, a 14 month old boy whose hyperphenylalaninemia had been detected during the routine Guthrie screening test. The only other unusual sign was that his blood levels of the vitamin folic acid were below normal. The low phenylalanine diet had been started in a timely manner, 12 days after birth, and blood phenylalanine levels had been kept within the recommended range. Although for the next seven months the baby appeared to be making normal progress, subsequently, the picture darkened. He often had seizures that could not be controlled by anticonvulsants. He had frequent drooling and eye-rolling and was unable to roll over or sit up without support. His electroencephalogram (EEG) was abnormal. All of these signs indicated that the patient had a serious neurological disease, somewhat resembling a parkinsonian state.

Tony wanted to know whether we could measure the amounts of phenylalanine hydroxylase, reductase and tetrahydrobiopterin in the patient's liver if he managed to obtain a liver biopsy sample from the patient and whether we could do it soon, very soon. He did not like the way the baby's condition was deteriorating.

The frozen liver sample arrived within a few days. The day after it came, we carried out the assays that we had developed for measuring the levels of the three components of the hydroxylating system. The results were unambiguous. The levels of the hydroxylase and tetrahydrobiopterin were in the normal range but there was not a trace of reductase (DHPR). This infant was the first documented case of PKU caused by a deficiency of the reductase. This was exciting not only because we had just discovered a new disease but also because it was one whose existence we had both predicted and feared.

As mentioned earlier, we had also predicted that because the reductase is essential for the synthesis of the neurotransmitters dopamine, noradrenaline and serotonin, patients lacking the reductase would be more seriously ill than patients with classical PKU who lack phenylalanine hydroxylase. A deficiency of these neurotransmitters would be disastrous because these small molecules are crucial for normal functioning of the brain and other parts of the nervous system. They are the chemical messengers that are released from a stimulated neuron and transmit a signal to adjacent neurons or to a target organ such as the heart. It is the way neurons "talk" to each other. There is also evidence that neurotransmitters may be essential for the normal development of the brain just as proper innervation is essential for normal development of muscle.

There seemed little reason to doubt that the neurological deterioration in this reductase- deficient patient was caused by lack of some or all of these neurotransmitters. Measurement of dopamine and serotonin levels in his brain and their metabolites in his CSF showed that their levels were very low compared to normal values. On the basis of these results, we recommended that patients lacking the reductase be treated with dopa, the precursor of the neurotransmitter dopamine, together with a drug called carbidopa (the combination is called "Sinemet" and is widely used in the treatment of Parkinson disease) and 5-hydroxytryptophan, the precursor of the neurotransmitter serotonin, as well as with a low phenylalanine diet to control their hyperphenylalaninemia. Although this treatment, started when the patient was 22 months old, normalized the low levels of dopamine and serotonin and appeared to slow down the neurological deterioration, this delayed start of the treatment proved to be crucial. Sadly, within the year the patient died. Nonetheless, we were still confident that this treatment would work when started early enough and that its failure with this particular patient was due to the unavoidable long delay in starting the treatment. The experience provided another example of the truth of the nursery rhyme about how difficult it was to put Humpty Dumpty together again after his fall. It emphasizes the fact that it is easier to try to prevent a disease than to successfully treat it once it has taken hold.

The reductase, DHPR, is an indispensable part of the enzyme systems that convert phenylalanine to tyrosine and of those that synthesize the essential neurotransmitters dopamine, noradrenaline and serotonin. DHPR is needed because of the way that tetrahydrobiopterin functions in these systems. In the phenylalanine hydroxylating system, for instance, for each molecule of phenylalanine converted to tyrosine, a molecule of tetrahydrobiopterin gets used up (by a process known as oxidation). In order for the conversion of phenylalanine to tyrosine to proceed, the tetrahydrobiopterin must be continuously regenerated (by a process known as reduction). It is the job of DHPR to carry out this latter process. The actual reducing agent used in this reaction is another coenzyme, called NADH, which, in turn, gets its reducing power mainly from the sugar that we eat. The same cycle of oxidation and reduction of the pterin coenzyme takes place during the synthesis of some of the neurotransmitters.

An understanding of the way the reductase works explains why trying to treat DHPR deficiency by giving tetrahydrobiopterin to these patients is not feasible. The amount needed would be equal to the amount of phenylalanine converted to tyrosine each day, approximately 100 milligrams per kilogram of body weight each day. At current prices, this treatment would cost about $58,000 per year for a 20 pound infant and almost $300,000 per year for a 100 pound youngster.

The confidence we had in the effectiveness of the treatment of reductase deficiency with neurotransmitter precursors was ultimately vindicated. It has become the standard therapy for this disease. But like the low phenylalanine diet treatment of classical PKU, it must be started as soon as possible after birth for the maximum benefit. It may be recalled that Tony Holtzman's patient had low blood folic acid (also called folate) levels. In my laboratory, we showed that DHPR plays a role in maintaining normal tissue levels of folic acid, a finding that could explain why folate levels were below normal in Tony's patient. Related to these results, it has been found that in some reductase- deficient patients, administration of a form of folic acid such as folinic acid or 5-methyltetrahydrofolate greatly improves the outcome over that seen with neurotransmitter precursors alone.

Given the importance of early treatment of reductase deficiency, early diagnosis is imperative. Fortunately, unlike the situation with classical PKU, where the affected enzyme, phenylalanine hydroxylase, is present only in liver, DHPR is widely distributed. Of particular importance for diagnostic purposes, It is found in white and red blood cells and is still detectable in a sample of blood dried on a piece of filter paper. This enables samples to be sent by mail to medical centers where DHPR activity can be measured. Routine screening of newborns for PKU now includes the Guthrie assay for blood phenylalanine levels as well as the assay for DHPR activity, both carried out on a drop of dried blood on a piece of paper. This variant form of PKU is extremely rare, occurring in around 1 in 500,000 births.

In the months following the publication in the New England Journal of Medicine of our report describing this first documented case of DHPR deficiency, I received a steady trickle of inquiries from clinicians, mainly pediatricians, describing other non-responsive PKU patients. We carried out assays of the components of the phenylalanine hydroxylating system on liver biopsy samples from some of these patients and found 4 additional cases of DHPR deficiency. Other cases were being detected and reported from universities in Europe, Australia and Japan.

Hyperphenylalaninemia caused by a deficiency of tetrahydrobiopterin

Sometime in 1977, I was contacted by Dr. Stan Berlow at the University of Wisconsin about another PKU patient, a 4 year old boy, Tim D., who was not doing well despite having been started on a phenylalanine- restricted diet at the age of 25 days with good subsequent control of his blood phenylalanine levels. Although he did not have a history of seizures, this patient also showed signs of neurological disease. At the age of 6 , he could walk but only with support and even then for only very brief periods; his primary way of moving around was by crawling. He was described as dystonic and spastic and he drooled a lot. His speech was slurred and was judged to be at the level of a 2 or 3 year old.

Dr. Berlow sent us a small frozen piece of the patient's liver obtained by open biopsy to see if we could determine which component of the phenylalanine hydroxylating was deficient. We quickly found that levels of phenylalanine hydroxylase and DHPR were normal, whereas the level of the coenzyme, tetrahydrobiopterin, was extremely low, only 7% of the normal value; a similar deficit of the coenzyme was found in urine and CSF. Our results showed that at the molecular level this patient's disease was not the same as that of Tony Holtzman's patient. This was the first patient described who was suffering from a variant form of PKU due to a documented hepatic deficiency of tetrahydrobiopterin.

Later, we showed that the deficiency of tetrahydrobiopterin in this patient was due to an inability to synthesize it. We also pinpointed the step that was blocked by showing that neopterin, a precursor of biopterin, accumulates in the urine and liver of the patient, indicating that the blocked step lies between neopterin and biopterin, a step catalyzed by an enzyme bearing the ungainly name pyruvoyltetrahydropterin synthase (PTPS). Since tetrahydrobiopterin, like DHPR, is also essential for the synthesis of the neurotransmitters, dopamine, noradrenaline and serotonin, it was not surprising that this patient was found to be deficient in all of these compounds that are so vital to the normal development and functioning of the brain. This variant form of PKU can be diagnosed by a determination of the amount of neopterin and biopterin in urine. It is twice as common as DHPR deficiency.

In contrast to DHPR-deficient patients, where, as has already been discussed, successful treatment with tetrahydrobiopterin would require that enormous amounts be given every day, the prospects for therapy with tetrahydrobiopterin being used successfully in patients lacking PTPS, such as Tim, were propitious. Since they have a normal complement of DHPR, they should be able to regenerate the coenzyme as it is used up. This made it likely that much smaller doses of tetrahydrobiopterin would be needed.

There was, however, a problem with the use of tetrahydrobiopterin to treat this illness: this compound was reported to be unable to enter the brain because it could not cross the "blood-brain barrier", a protective shield of cells that surrounds the brain and restricts the entry of certain substances, including some very useful drugs like tetracyclines and sulfonamides, as well as some sugars and amino acids.

A possible way to overcome this obstacle was to find a substitute for tetrahydrobiopterin that had coenzyme activity with the hydroxylases that would also be able to get into the brain. Since the blood-brain barrier was known to be particularly good at excluding molecules that were not very soluble in fats, the way to trick the barrier was to find a substitute that is more fat soluble than tetrahydrobiopterin.

I knew what seemed to be an ideal candidate. It was 6-methyltetrahydropterin, an analogue of tetrahydrobiopterin that I had synthesized years before and had shown to have high coenzyme activity with the hydroxylases. Indeed, in the test tube, it was at least as active as tetrahydrobiopterin . It was also more fat soluble than tetrahydrobiopterin and was therefore likely to be able to penetrate the barrier better than the naturally-occurring coenzyme.

We carried out a preliminary experiment in rats comparing the ability of 6-methyltetrahydropterin and tetrahydrobiopterin to enter the brain. As expected, the methyl compound did cross the blood-brain barrier with ease, about 10 times better than tetrahydrobiopterin. Unexpectedly, however, at the dose used in the experiment, very substantial amounts of the latter compound also got into the brain, enough to double the amount that was already there. These results dispelled the prevailing dogma about the inability of tetrahydrobiopterin to enter the brain. They also showed that the methyl analogue had therapeutic potential.

We were interested in carrying out a trial treatment with tetrahydrobiopterin on Tim as soon as possible. I raised the possibility of carrying out the study at NIH with Dr. Joe

Schulman, a friend and a physician/scientist who was working at the National Institute of Child Health and Human Development. He thought that it would be possible and agreed to serve as the admitting and attending physician.

Once all of the considerable paper work needed to admit him had been completed, Tim was brought to NIH by his parents. As soon as he was settled in his room in the Clinical Center, the hospital on the NIH "campus", Joe Schulman and I went over to meet them there. I was, of course, well aware of Tim's very serious neurological deficits from Dr Berlow's description, but I was not prepared for what I saw. As I entered the room, he was just returning to his bed from the toilet. He displayed the full repertoire of his physical symptoms: he appeared to be unable to walk without his mother's help. Even with her help, his legs looked as if they were made of rubber. In addition, he drooled continuously and had very frequent eye-rolling. As soon as he was back in his bed, he showed another characteristic of his illness, profound lethargy.

One look at Tim convinced us that it would be a good idea to start the treatment with tetrahydrobiopterin as soon as possible. As far as the question of what dose to use, we were guided by the results of our earlier study with rats that showed that after a dose of tetrahydrobiopterin of 20 milligrams per kilogram of body weight, considerable amounts of the administered dose had gotten into the brain. We decided to start treating Tim with the same dose, given orally every 12 hours at 10 milligrams per kilogram. Blood and CSF samples were to be collected a few hours after the dose had been given to measure how much of the compound had gotten into the brain and whether the treatment had any effect on the depressed levels of neurotransmitters.

I slept very badly that night. For one thing, I was worried about possible serious side effects of giving the coenzyme at such a large dose. Even though it is a naturally-occurring compound, nobody had ever been given this amount. Ironically, I was also too excited to sleep because of the possibility that this treatment might work.

The next morning, there was a written telephone message waiting for me at my office from Bill Rizzo, one of Joe's young assistants, a physician who was helping in various aspects of the experiment. The actual wording of the message was:

"100% better! Amazing. Muscle strength, coordination poor, couldn't walk, eyes rolling, speech slurred. All corrected!!"

I dashed over to the Clinical Center to see for myself. I was surprised and, for a moment, concerned to find that Tim's room was empty. The nurse told me that he was in the lounge down the hall. I could not believe what I saw. Tim was sitting at a table working on a jigsaw puzzle, picking up the small pieces without any apparent difficulty. His speech, although not quite normal, was now easily comprehensible. His parents were beaming. We were all ecstatic.

Later, the results of the analyses that had been carried out on Tim's blood and CSF samples were available. They were coherent with his improved behavior and with what we had seen with our own eyes. Levels of tetrahydrobiopterin in both plasma and CSF were markedly elevated. Although levels of his previously depressed neurotransmitters and their metabolites were also increased, they were still below normal. These results strongly suggested that these increases were responsible for his improved behavior. During the same hospitalization, we carried out a separate trial using a similar dose of the synthetic coenzyme, 6-methyltetrahydropterin so that we could compare its effectiveness with that of tetrahydrobiopterin. The improvement in behavior and in his neurotransmitter levels was just as dramatic as it was with the naturally-occurring coenzyme. This was an exciting result because the 6-methyl compound was much cheaper than tetrahydrobiopterin.

Because treatment with neither tetrahydrobiopterin nor the methyl compound increased Tim's neurotransmitter levels to normal, dopa and 5-hydroxytryptophan, the precursors of these neurotransmitters were added to his treatment regimen. He has now been treated with this combination for 17 years, longer than any other patient with this disease. His improvement has been sustained during this whole period. Subsequent to our original study, we also initiated what we thought would be a long-term therapeutic trial in which we substituted the cheaper 6-methyltetrahydropterin for tetrahydrobiopterin. That trial, however, was promptly terminated when there were early signs that this compound might be toxic to the liver. Its great therapeutic potential, therefore, was never realized.

The exhilaration that we all felt after the successful treatment of Tim with tetrahydrobiopterin did not last too long. Shortly after this stunning success, we carried out a similar therapeutic trial on another tetrahydrobiopterin-deficient patient but with very different results. Our preliminary studies showed that this little girl, Rosaria, a patient of Dr. Rod McInnes at the Hospital for Sick Children in Toronto, had a more severe deficit of tetrahydrobiopterin than Tim. Despite the fact that this deficit was corrected by the administration of the tetrahydropterin, there was only modest improvement in her neurological signs. Correlating with this lack of success and, perhaps responsible for it, there was also no normalization of the very low levels of neurotransmitters in her CSF. There was a somewhat better response to treatment with the neurotransmitter precursors dopa and 5-hydroxytryptophan.

These 2 patients illustrate the extremes seen in treating this variant form of PKU (PTPS deficiency). With Tim, treatment with tetrahydrobiopterin alone was remarkably successful in improving his neurological deficits. With Rosaria, this treatment was almost totally ineffective and, even in combination with dopa and 5-hydroxytryptophan, was only modestly successful.

Most PTPS-deficient patients fall between these 2 extremes. For them, combined therapy with tetrahydrobiopterin plus neurotransmitter precursors is at least partially successful. Combined therapy has become the standard treatment for this disease.

Finding the reason for the variable response is a challenging problem for future research. A clue may be provided by one of the obvious differences between Tim and Rosaria: the latter child had a much more severe deficit of tetrahydrobiopterin in her cerebrospinal fluid than did Tim. It may be that a certain minimum amount of tetrahydrobiopterin is needed for normal fetal brain development, especially for normal neuronal growth. In order for tetrahydrobiopterin to be therapeutically effective, it must be able to stimulate the activity of tyrosine and tryptophan hydroxylases so that they can make more of their essential neurotransmitter products. Since in the brain these enzymes are located within neurons, normal neuronal growth with a normal intraneuronal compliment of these hydroxylases, may be a prerequisite for the fully successful treatment of this disease with tetrahydrobiopterin. This notion also fits in with the fact that Rod's patient was about 1 month premature; her brain, therefore, would have been deprived of the protective effect of maternally-derived tetrahydrobiopterin for this period.

This scenario raises the possibility that a higher rate of success could be achieved with these patients if tetrahydrobiopterin was given to the mother prior to the birth of the affected baby. To carry out this kind of prenatal treatment, however, it would be necessary to carry out prenatal diagnosis of the disease. Although this is feasible, it is not practical. Given the rarity of the disease, the prospects for widespread prenatal diagnosis are extremely remote. To pursue this particular line, perhaps the best course that can be considered at the present time is a compromise that could be used in those families where a child with PTPS- deficiency has already been born. If a family like this were considering having another child, prenatal diagnosis would make sense and, if the results indicated that the fetus were PTPS-deficient, the parents could be made aware of the possible beneficial effects of prenatal treatment with tetrahydrobiopterin. Right now, the only practical way to improve the outcome of treatment of this disease is to have the earliest possible postnatal diagnosis and initiation of combined therapy with tetrahydrobiopterin and neurotransmitter precursors.

The notion that serotonin may be crucial for normal brain development and functioning is supported by results of animal studies that showed that depletion of serotonin in the immature brain leads to a long-lasting loss in the number of synapses in the adult brain. Since synapses are essential for learning and memory processes, these findings would predict that animals with defects in synaptogenesis would be impaired in their ability to learn new tasks. Results of the performance of serotonin-depleted animals during maze testing confirmed that their ability to learn was seriously compromised. If these results can be extrapolated to humans, they could explain why attempts to treat PKU due to defects in tetrahydrobiopterin synthesis postnatally with the pterin cofactor together with the serotonin precursor 5-hydroxytryptophan and the dopamine precursor dopa have not been fully successful. Again, they point to the need to initiate such a treatment prenatally.

Why are these neurotransmitters so important for the developing brain? For serotonin, a point that must be emphasized in this regard is that serotonin functions in the brain not only as a neurotransmitter but also in a more ancient role of a growth-regulating or trophic factor. Indeed, it plays this other role, which includes its ability to regulate the development and maturation of the brain, before it assumes its role as a neurotransmitter. That it plays roles beyond that of a neurotransmitter could be surmised from its presence even in some one-cell organisms like protozoans, that are devoid of a nervous system.

A reflection of some of serotonin's many roles can be seen from the consequences of its removal: these include a decrease in the inhibition of aggression against oneself (suicide) as well as against others (homicide), changes in sexual behavior, learning, sensitivity to pain, attention, body temperature, appetite, steroid hormone secretion, respiratory rate, blood flow and sleep (its effects on sleep are so pronounced that it used to be called "somnotonin" and was thought to be the sleep neurotransmitter. This action of serotonin may well be the basis for the old folk remedy for sleeplessness, a glass of warm milk before bedtime. It may work because of the presence in milk of protein that contains a relatively high amount of tryptophan, the precursor of serotonin. It has been said that the functional consequences of removing serotonin impact nearly every behavioral and biological process studied. In addition to biological processes just mentioned, altered brain serotonin levels have been implicated in many psychiatric and neurological disorders including depression, schizophrenia, Down syndrome, Alzheimer disease, autism, alcoholism, attention deficit disorder and sleep apnea.

What, if anything, does this compilation of effects of serotonin deficiency have to do with PKU? The answer is that this list could serve as a road map that identifies potential medical problems that could become manifest in patients with serotonin deficiency secondary to a lack of tetrahydrobiopterin. Milder manifestations of these problems could also be seen in classical PKU patients.

Unfortunately, there is no indication that anyone has consulted this road map for such a purpose. At most, there have been sporadic reports of an occasional patient who is afflicted with one of the variant forms of PKU who has a disturbed sleeping pattern. Sometimes these patients sleep only for a total of 2 to 3 hours a night. In some of these patients, increasing the amounts of 5-hydroxytryptophan (a serotonin precursor), which is part of the accepted treatment for this disease, have been found to dramatically improve the sleep problem. In addition, there has been at least one report of a child with hyperphenylalaninemia caused by a defect in biopterin synthesis who had periods of elevated temperature and increased respiratory rate for which no cause could be found.

An explanation for the variable effectiveness of large doses of tetrahydrobiopterin in the treatment of variants due to defective synthesis of the pterin that has not been tested may be related to the ready oxidation of this compound by oxygen. During this reaction, potentially toxic compounds like hydrogen peroxide are formed. One of the main defenses that have evolved to protect organisms, including humans, from these toxic effects is an enzyme called glutathione peroxidase, which destroys hydrogen peroxide by accelerating its reaction with a small molecule known as glutathione. It has been found that the trace element selenium is an integral part of the peroxidase molecule and that it is essential for the enzyme's activity.

Whereas selenium deficiency is rare in humans who eat a diet typical of Western countries (it is seen in parts of China), it is common in PKU patients who are on a low-phenylalanine diet with restricted intake of natural protein. Consequently, the activity of glutathione peroxidase in red blood cells is decreased in these treated patients. It may be that tissue levels of the peroxidase are also low.

The ability of tetrahydrobiopterin to generate hydrogen peroxide, coupled with the diminished activity of glutathione peroxidase, one of the organism's main defenses against hydrogen peroxide toxicity, may limit the effectiveness, and contribute to the variable results of the tetrahydrobiopterin treatment of biopterin-deficient variants of PKU. To avoid this potential problem, the selenium status of these patients should be normalized prior to the initiation of treatment with tetrahydro-biopterin.

This discussion of the findings that even the combined therapy is not fully effective with every patient who suffers from a block in the synthesis of the coenzyme should not blur the bottom line, i.e. that the development of this therapy represents a stunning medical triumph over a disease that was once called "lethal" and "malignant". Several hundred children are alive today because of this therapy.

In 1984, Niederwieser and his colleagues in Switzerland described still another variant form of PKU due to a diminished ability to synthesize the coenzyme. These patients are blocked in the very first step in the pathway leading to the synthesis of tetrahydrobiopterin, a step catalyzed by an enzyme called GTP cyclohydrolase. In contrast to the patients who excrete a lot of neopterin but very little biopterin, these excrete neither neopterin nor biopterin. They can be easily distinguished from the others, therefore, by an analysis of these 2 compounds in the urine. This is the rarest of the variants, accounting for only about 5% of all of those who are unable to synthesize tetrahydrobiopterin. Their neurological deficits as well as their treatment are the same as those for PTPS-deficient patients.

Hyperphenylalaninemia caused by a deficiency of dehydratase (PHS).

With the demonstration that the phenylalanine hydroxylating system consists of three components-phenylalanine hydroxylase, dihydropteridine reductase, and tetrahydrobiopterin- our general understanding of how the system works seemed to be complete. There remained, however, a nagging doubt about just how complete this understanding was. The doubt came in the form of a protein, probably an enzyme, that we had discovered in the 1970's and purified from rat liver. Other than knowing that it could stimulate the enzymatic hydroxylation of phenylalanine, we had no idea about how it worked. Until its precise function was known, we referred to this protein or enzyme as "phenylalanine hydroxylase stimulator" or "PHS". The awareness of its existence made us feel uneasy. We felt like someone who had just completed a jig-saw puzzle only to realize that there was still an unused piece of the puzzle left over.

A clue to how PHS might function came from our observation that during the conversion of phenylalanine to tyrosine, in the absence of PHS, tetrahydrobiopterin was converted to another pterin, one bearing an extra hydroxyl group This new pterin was rather unstable; at neutral pH it slowly broke down to dihydrobiopterin. Significantly, PHS was able to stimulate this breakdown. We proposed and later proved that PHS was an enzyme, a pterin dehydratase and that the new pterin was an intermediate between tetrahydrobiopterin and dihydrobiopterin. Furthermore, PHS was essential for the regeneration of tetrahydrobiopterin from this intermediate. The ability of the PHS-catalyzed reaction to occur slowly even in the complete absence of PHS, especially at neutral pH, probably accounts for why this enzyme was not discovered earlier.

The discovery of PHS raised the question of whether a variant form of PKU might exist caused by a deficiency of PHS. Because the PHS-catalyzed reaction can occur slowly even in the absence of PHS, at least in the test tube, we predicted that a total lack of PHS would probably result in only mild hyperphenylalaninemia. For several decades, we were in the strange position of having identified an enzyme that could theoretically cause PKU or hyperphenylalaninemia, waiting for the appearance of a disease caused by the lack of this enzyme. As will be discussed in a later section, that wait came to an end about 20 years later with the description, as we had predicted, of a variant form of PKU with mild hyperphenylalaninemia caused by a lack of PHS.

In 1988, several patients with modestly elevated blood phenylalanine levels were described who appeared to be prime candidates for PHS deficiency, if for no other reason than that they both presented with a unique metabolic symptom: they both excreted a novel pterin. Curtius and his colleagues in Switzerland identified the novel pterin as 7-biopterin, a rearranged or isomeric form of biopterin. (Two compounds that are isomers bear the same relationship to each other as two rooms of the same dimension, both filled with identical pieces of furniture but with the furniture arranged differently in each room).

The Swiss workers also showed that the excretion of 7-biopterin was increased when the patients were fed tetrahydrobiopterin, a strong indication that the 7-isomer was formed in the body from tetrahydrobiopterin. That the formation of 7-biopterin might be consequent to a lack of PHS was strongly suggested by results of test tube experiments, carried out independently in my lab and in Curtius' lab, showing that during the hydroxylation of phenylalanine by the hydroxylating system in the absence of PHS, a small amount of tetrahydrobiopterin was converted to 7-biopterin. These results consolidated the idea that excretion of 7-biopterin is the hallmark of PHS deficiency; in other words, that excretion of 7-biopterin is a marker for this variant form of PKU. We also showed that the tetrahydro derivative of 7-biopterin was a potent inhibitor of phenylalanine hydroxylation. It could therefore slow down the metabolism of phenylalanine.

The great remaining challenge was to prove that patients who excrete the 7 isomer are actually deficient in PHS. Unfortunately, the approach that Jervis and I had used to prove that classical PKU is caused by a lack of a functioning phenylalanine hydroxylating system- to show that the system is not active in a liver biopsy sample obtained from a PKU patient- was no longer an option. The research climate in the United States had changed so that it was almost impossible to get a fresh liver biopsy sample. A further limitation to the options available to us was the narrow tissue distribution of PHS. This enzyme was not present in readily available tissues like skin or blood cells. Given these limitations, how could the structure of PHS or its enzymatic activity in patients be determined?

A few years ago, there would have been no way to address this question. We would have reached the end of the line with this research problem. Luckily, the relatively new field of molecular biology provided a pathway around around the block. We didn't know it at the time but the path that we chose to pursue was fateful and would lead us to a hidden treasure that would reveal facets of the PKU/hydroxylase story that were previously not only unknown but were unimaginable.

Utilizing the amino acid sequence of rat liver PHS and the techniques of molecular biology, we were able to determine

119

the DNA sequence of the enzyme. Once that step had been accomplished, we set out to compare our sequence with those of every other protein whose DNA sequence has been worked out. Fortunately, in the computer age this comparison takes only a few minutes because these sequences are stored in computer "data banks". To our amazement, we quickly found that an essentially identical DNA sequence was already in the data bank. But it was not listed as the sequence of PHS/dehyratase but rather as the sequence corresponding to a protein I had never even heard of, called DCoH. This protein, which had been discovered by Gerry Crabtree and his colleagues at Stanford University, was known to be involved in the regulation of the transcription of a large family of hepatic genes. Since gene transcription is an essential step in the synthesis of proteins within the cell, DCoH was an important regulator of the amount of certain proteins that are synthesized in the liver.

When I telephoned Crabtree to tell him what we had found, his initial response was to raise the possibility that our preparations of PHS had been contaminated with some DCoH protein and what we had actually sequenced was this contaminant. We both agreed that the way to rule out this possibility was for us to exchange proteins and test each one for both activities. The results of these tests were utterly clearcut. Crabtree's preparation of DCoH had high PHS/dehydratase activity-comparable to our pure PHS fractions when tested with the phenylalanine hydroxylating system- and our PHS preparation had high DCoH activity in his gene transcription system. These results came close to proving that a single protein molecule had two disparate biological activities. This phenomenon had already been described with other proteins. It was called "molecular opportunism".

After we had digested this astonishing finding, we turned our attention back to the original question- are hyperphenylalaninemic patients who excrete 7-biopterin deficient in PHS/dehydratase? We succeeded in locating a candidate patient, a seemingly healthy one-year-old boy with mild hyperphenylalaninemia whose metabolic abnormalities had been picked up during routine newborn screening. From samples of the patient's and his parent's blood, we isolated genomic DNA which contained the DNA sequence corresponding to PHS. We showed that our hunch was correct. The PHS gene from the patient harbored two mutations, one inherited from his father and one from his mother. We postulated that these mutations adversely affected his PHS activity and led to his mild hyperphenylalaninemia. A few years later, we proved that this postulate was correct. Using the techniques of molecular biology, we constructed the two mutant forms of PHS/dehydratase and showed that, as predicted, the combined effect of these mutations was a marked (but not a total) decrease in dehydratase activity. This work was confirmed by the Swiss workers.

To our great surprise, these two mutations that so crippled the dehydratase activity of PHS had little effect on its gene transcriptional or DCoH activity. We did find, however, that the gene for phenylalanine hydroxylase was one of those that may be regulated by the DCoH activity of the protein. These findings raise the possibility that a new form of PKU may be lurking out there caused by mutations in the DCoH/ PHS gene.

Our results indicate that the PHS/dehyratase- deficient patient that we have studied has significant residual dehydratase activity. It is possible that patients who are totally devoid of this activity will have a more virulent form of hyperphenylalaninemia or PKU. If such patients exist, the studies described above on a mildly affected patient with some residual dehydratase activity should be useful in dealing with the more serious versions of the disease. Indeed, knowing that PHS functions with the hydroxylaing system by limiting the amount of tetrahydrobiopterin that is converted into 7-biopterin suggests that such patients could probably be successfully treated by the administration of extra tetrahydrobiopterin.

Tetrahydrobiopterin-responsive PKU

Recently, a new variant form of PKU was reported by Shigeo Kure and his colleagues in Japan: tetrahydrobiopterin-responsive phenylalanine hydroxylase deficiency. These patients were considered to have mild hyperphenylalaninemia because their serum phenylalanine levels never exceeded 20mg/dL even when they were not on a phenylalanine-restricted diet. The distinctive feature of this disease was revealed, when, as part of a differential diagnosis to detect a possible deficiency of tetrahydrobiopterin, these patients were given a large oral dose of tetrahydrobiopterin (10 mg/kg body weight) after their blood phenylalanine levels were first allowed to rise in response to feeding them a diet containing phenylalanine for a few days. It was found that the serum levels of phenylalanine gradually decreased following the dose of tetrahydrobiopterin. Phenylalanine levels approached, but did not reach, normal levels even after 24 hours. Later studies of similar patients by Dutch workers showed that blood levels of phenylalanine were actually normalized after higher oral doses (20mg/kg body weight) of tetrahydrobiopterin.

At first glance, these results might have suggested that these patients were suffering from a variant form of hyperphenylalaninemia caused by a lack of tetrahydrobiopterin. It may be recalled that patients with classical PKU, because they lack any functional phenylalanine hydroxylase, do not show any decrease in serum phenylalanine levels in response to a dose of tetrahydrobiopterin. In contrast, patients whose hyperphenylalaninemia is due to defects in tetrahydrobiopterin synthesis or regeneration show this response to the pterin cofactor. The possibility that these new patients lacked tetrahydrobiopterin, however, was conclusively ruled out by the finding that their urinary levels of biopterin and its metabolites were normal. In addition, activities of dihydropteridine reductase were shown to be normal. Also speaking against the possibility that their hyperphenylalaninemia was caused by any defect in synthesis or regeneration of the pterin cofactor was the finding by the Dutch workers that CSF levels of metabolites of amine neurotransmitters, such as serotonin, norepinephrine and dopamine were normal.

That the villain in these cases was phenylalanine hydroxylase was strongly indicated by the detection of mutations in the hydroxylase gene. The finding that the hyperphenylalaninemia could be relieved by increasing the levels of tetrahydrobiopterin pointed to the likelihood that these particular mutations resulted in a decreased affinity of the hydroxylase for the pterin cofactor. It has been suggested that patients with this novel form of hyperphenylalanine can be successfully treated with supplements of tetrahydrobiopterin .

CHAPTER 9

POTENTIAL PROBLEMS OF ASPARTAME

The discovery of the artificial sweetener, aspartame, also known as NutraSweet and Equal, was a boon to dieters and a tease to PKU patients. For the patients, it was forbidden fruit because this substance is a derivative of phenylalanine which breaks down to phenylalanine in the body. Beverages sweetened with aspartame such as Diet Coke and Kool Aid contain as much as 180 milligrams of phenylalanine in each 12 ounce serving, or one-third of the daily allowance of the amino acid for a PKU patient. This represents enough of a hazard for these patients that every blue packet of Equal displays the following warning: "PHENYLKETONURICS: CONTAINS PHENYLALANINE".

There is, however, no warning for pregnant women who happen to be PKU heterozygotes. The question of whether they should be warned has been debated for years but not yet answered. The answer depends on how likely it is that a pregnant PKU heterozygote could consume enough aspartame to damage her fetus. We know from the condition called maternal PKU (reviewed in an earlier chapter) that a sufficiently high level of phenylalanine in the mother's blood will play havoc with the developing heart and brain of the fetus.

Whereas the experience with maternal PKU tells us loudly and clearly that high levels of phenylalanine are damaging, it is not as informative about only moderately elevated levels. Indeed, on this point it mumbles and stutters. There were early indications that fetal damage only occurs when the mother's blood phenylalanine levels exceed a threshold value that is 10 times higher than the normal (i.e, 0.6mM). This was a comforting result because it was inconceivable that eating aspartame could ever increase blood phenylalanine to that level.

As mentioned earlier, however, later work challenged the "threshold theory". Specifically, some of this work showed that damage can occur when the mother's blood phenylalanine levels are only 5 to 6 times higher than normal (0.3-0.4mM). Indeed, Dr. Isobel Smith in 1987 published data showing that for each 200 μM increment in the level of phenylalanine there is a 0.5 cm deficit in head circumference size of the newborn from the mean value of 35 cm. If these results are accepted, the question that must be addressed is what are the chances that any pregnant woman would ingest enough aspartame-sweetened drinks to increase her blood phenylalanine to dangerous levels?

I used to think that the chances that this could happen were as remote as winning the lottery. But a few years ago I received a phone call from a pregnant woman that changed my mind. She wanted to know whether I thought it was safe for her to be drinking things like "diet coke". I said that there was probably no harm in drinking an occasional aspartame-sweetened soda. What did I mean by "occasional"? I stuck my neck way out and said that 1 can a day was probably OK. How about a 6-pack (i.e., 6 cans, 12 ounces each) every afternoon? That's the amount that she had been drinking. Like a startled turtle, I withdrew my neck. No, I didn't think that was a good idea and that if she were a PKU heterozygote, the idea would be twice as bad. She, of course, did not know whether she was a PKU heterozygote.

The problem here is that a precise answer to the question posed by that lady is not known. But results such as those mentioned above on the deleterious effects on the baby's head size of relatively small increases in phenylalanine blood levels are worrisome. Nonetheless, since the odds against any individual being a PKU carrier are about 50 to 1, a pregnant women without any history of PKU in her family could probably drink a few cans a day of an aspartame-sweetened drink with very little risk to her unborn baby.

CHAPTER 10
PROSPECTS FOR ALTERNATE THERAPY
FOR PKU

Although the vast majority of PKU families cope amazingly well with the burdens of the dietary treatment for the disease, there can be little doubt that most of them hope and pray for a different treatment. A recent issue of National PKU News, a publication devoted to promoting the exchange of information about PKU, expressed this hope succinctly in a lead article that started out as follows: "We can remain optimistic that someday in the not-too-distant future there will be a true 'cure' for PKU, making the diet no longer necessary."

How realistic is this optimism? For many years, the hopes for cures for genetic diseases, in general, were too high. They were driven, in part, by the unwarranted assumption that has been repeated so often in the popular press that it is now axiomatic: The identification of an affected gene (or genes) in a disease will automatically lead to an effective treatment for that disease. Indeed, this notion seems so compelling and self-evident that a highly respected science writer recently fell into this trap and innocently rewrote the history of PKU. Specifically, he reported that the dietary treatment for PKU was the result of the demonstration that mutations in phenylalanine hydroxylase

128

cause PKU. As has already been discussed, however, the dietary treatment for PKU preceded by more than 30 years any knowledge about the molecular basis of the disease.

In the march to find a cure for genetic diseases, the perennial champion, the standard- bearer, has been the notion of somatic gene therapy, i.e., the idea that genetic diseases can be cured by administering the missing or the normal gene to a patient. That notion fits in with the revolution in medicine that took place in the middle of the 20th century that made it possible to think seriously about replacing worn out or defective organs and joints- hearts, kidneys, hips, knees- in our bodies. If a failing heart can be replaced, why not a defective gene?

The history of attempts to cure genetic diseases with gene therapy, unfortunately, falls into the category of one of Xeno's paradoxes, the one that describes the imaginary problem that you face if you are trying to get from point A to point B. If in the first time period you cover half of the necessary distance from A to B, and in the next period you cover half of the distance that still remains, and so on, the paradox is that you can never arrive at B, your destination, because you will always have that last remaining half of the distance to cover.

At the present time, this paradox comes close to characterizing the status of gene therapy. Progress in this field continues to be good, continues to cover half of the remaining distance to the goal of an effective treatment for a human genetic disease, but the final goal continues to be elusive.

The first step required for Gene therapy is the isolation of the normal gene for the enzyme that is affected in the disease of interest. For PKU, that step was accomplished in 1985, when Savio Woo first isolated the gene for normal human phenylalanine hydroxylase. The next step, the delivery of the gene into the liver of a PKU patient, has not yet been accomplished. It is a daunting task. With very few exceptions, simply mixing genes with cells does not work. What is needed is some kind of delivery system, a vector, for carrying the gene and depositing it in the cells where it will do its work of correcting the metabolic lesion.

The vectors that have received the most attention for treating genetic disease are viruses. These cellular parasites cannot reproduce by themselves. Through millions of years of evolution, however, they have been highly selected for their ability to attach themselves to the outer membrane of cells and to transfer their own genes into these cells. Once inside, the viral genome can hijack the protein synthesizing machinery of the host cell and use it to direct the synthesis of its own viral proteins. The result? The host gets sick, from a disease like influenza or polio, depending on the particular virus that is doing the hijacking.

Almost 40 years ago, scientists began to realize that it might be possible to use some biological jujitsu on viruses in order to exploit their ability to penetrate cells for the purpose of gene therapy. In the 1960's, Stanfield Rogers, a scientist working at the time at the University of Tennessee and later at the Oak Ridge National Laboratories, made a chance observation that demonstrated the potential of viruses as vectors for gene transfer into animal cells. He was working with the Shope papilloma virus that causes warts and observed that the skin of rabbits that were infected with this virus had high activity of arginase, an enzyme that breaks down the amino acid arginine, which like phenylalanine, is a normal component of proteins. Although liver tissue has a lot of arginase, skin from uninfected rabbits was known to be devoid of this activity. He also showed that rabbits infected with the Shope virus had very low blood levels of the amino acid arginine, presumably due its increased breakdown by the elevated amounts of arginase. The indications were that gene transfer had taken place from the virus to the rabbits and that viral genes had commandeered the infected rabbit's protein- synthesisizing machinery and were dictating that it make more arginase.

Rodgers wondered whether people in his lab who had been handling the virus-infected rabbits also had low blood levels of arginine. He found that they did. These findings indicated that the virus had succeeded in transferring to the lab workers at least some of its genes, including the arginase gene, and that this gene was being expressed and was functional. It was increasing a normal metabolic process, breakdown or catabolism of arginine in these people. Rodgers had accidently demonstrated the feasibility of the use of a viral vector to deliver a gene to humans.

One of the types of viruses that has been intensively studied for possible use in gene therapy are the adenoviruses, the infectious agents that cause mild flu-like symptoms in humans. Before they can be used as vectors, the viral genes responsible for such symptoms are crippled and the desired human gene (e.g., the gene for normal human phenylalanine hydroxylase) is inserted into the viral DNA. The product of this insertion is a new species of DNA, dubbed "recombinant DNA", because it is the result of the combination of, in this case, DNA from the hydroxylase with the DNA of the virus. It is an artificial gene that does not exist in nature.

The early experience with this particular species of recombinant DNA illustrates both the power and the current limitations of the use of viruses as vectors for gene therapy. An opportunity to explore the effectiveness of this vector for treatment of PKU presented itself in the early 1990's with the availability of a high-fidelity mouse model for the disease, developed by Dove and his colleagues at the University of Wisconsin. They produced mice lacking phenylalanine hydroxylase by treating them with a mutagen. This treatment introduced mutations in the hydroxylase gene that resulted in the production of inactive hydroxylase and consequent profound hyperphenylalaninemia in the treated animals. Because the mutation affected the gene in germ cells, the mutation was inherited in the offspring from the treated mice, opening the way to development of a strain of PKU mice which has been invaluable for research on this disease.

As a preliminary step to trying to treat PKU patients with gene therapy, Savio Woo and his colleagues first tried to "cure" Dove's PKU mice by the same means. They constructed a recombinant adenoviral vector containing the human phenylalanine hydroxylase gene and introduced it into the portal vein of the mutant mice. Infused by this route, the hydroxylase gene was most likely to reach the liver, the ideal target since it contains the full complement of the reductase and tetrahydrobiopterin, the other components needed to reconstitute an active hydroxylating system.

The treatment at first looked like it was a total success. Within a week, the elevated blood levels of phenylalanine in the treated mice had returned to normal. Unfortunately, however, the beneficial effect only lasted about 2 to 3 weeks. After that time, phenylalanine levels soared to their original values. Even worse, a repeat infusion of the relapsed mice with a second dose of the adenovirus-hydroxylase construct proved to be totally ineffective. The most likely reason for this failure was that the mice had developed antibodies against the adenoviral vector which were capable of fighting off this viral infection and short-circuiting its potential benefits. Before this approach can be tried for treating PKU in humans, a way must be found to prevent the immune response from sabotaging the treatment. This may well happen within the next few decades.

On paper, at least, an attractive alternative to somatic gene therapy for a disease like PKU is enzyme replacement therapy. The enzyme of choice obviously would be phenylalanine hydroxylase. Not only would it dispose of potentially damaging excess phenylalanine, but it would also provide tyrosine. Since healthy people fill their needs for this amino acid partly from the food that they eat and partly from the hydroxylation of phenylalanine, levels of tyrosine tend to be rather low in PKU patients because they do not get any from the latter source.

Several intimidating obstacles must be overcome before progress can be made in replacing *any* enzyme in humans; for the hydroxylase there are unique problems that are especially troublesome. Since all enzymes are proteins, the general problem inherent in any enzyme replacement is how to prevent the body's digestive enzymes in the stomach and the intestines from digesting and inactivating the enzyme that is being introduced. The special problem intrinsic to phenylalanine hydroxylase is that it is notoriously unstable in the test tube.

Given the obstacles to success, it may not be surprising that there have been pitifully few published studies in this area. In fact, the only one that I am aware of, published about 22 years ago, was inconsequential. Realizing that the hydroxylase had to be protected from destruction by digestive enzymes, the authors tried to encapsulate it in a synthetic porous gel made of acrylamide. This proved to be a poor choice since the acrylamide completely inactivated the enzyme in 10 minutes at room temperature. At body temperatures inactivation would have been almost instantaneous. This attempt at enzyme replacement was therefore aborted. Perhaps because of this negative experience, the possibility of treating PKU with exogenous phenylalanine hydroxylase remains largely unexplored.

Fortunately, even though phenylalanine hydroxylase might theoretically be the ideal choice for enzyme replacement therapy, use of other phenylalanine-metabolizing enzymes are being explored. Indeed, very promising results in "curing" hyperphenylalaninemia in the PKU mouse model have been obtained by Scriver and his colleagues in Montreal, Canada, with an enzyme called phenylalanine ammonia lyase (PAL). This plant enzyme doesn't convert phenylalanine to tyrosine, but, instead, breaks it down to ammonia and a substance called trans-cinnamic acid that is generally believed to be harmless. Both of these products are further metabolized to compounds that are rapidly excreted in the urine.

Compared to the hydroxylase, PAL has several important advantages: it is a more stable enzyme and it needs neither a coenzyme nor an ancillary enzyme in order to function. To keep the PAL from being digested, the enzyme has been fed to PKU mice in some kind of protected form. In one method, PAL has been immobilized in semipermeable microcapsules that keep the digestive enzymes of the mouse but not small molecules like phenylalanine from getting at the PAL, thus allowing the PAL to degrade the phenylalanine. In another method, naked, unprotected PAL has been fed to the mice together with a compound that inhibits the activity of proteolytic enzymes present in the intestine that would otherwise digest and inactivate the PAL. The logic behind feeding the encapsulated lyase is that it provides an alternative way to remove phenylalanine from the diet of a PKU patient. Instead of removing it from food in a processing factory, the way a product like lofenalac is manufactured, the encapsulated lyase, lodged temporarily in the gut, disposes of the phenylalanine as it is liberated from the food that is eaten.

Administered either way, the PAL was able to reduce by 40-50% the elevated blood phenylalanine levels in the PKU mice. This would probably not be enough to substitute completely for the low-phenylalanine diet in patients with classical PKU but it would be enough to allow for a considerable liberalization of it. That would represent a huge improvement in the life style of PKU patients and their families. Understandably, the PKU community is bubbling with excitement over these promising findings, even if they are in a preliminary stage. To see if this promise will be fulfilled, it is essential that long-term trials be carried out, first on PKU mice and then, of course, on PKU patients. Undoubtedly, such studies are ongoing.

In addition to attempts to treat PKU by introducing the gene for phenylalanine hydroxylase or another enzyme that can dispose of phenylalanine such as the lyase, a few proposals are on the drawing board that are aimed at a different kind of dietary treatment.

One of these alternatives stems from what is probably the most attractive theory about how high blood levels of phenylalanine damage the developing brain, the theory that proposes that excess phenylalanine crowds out other amino acids which are essential for normal brain development as they try to enter the brain. If that is, in fact, the mechanism by which too much phenylalanine causes mental retardation, it prompts the following question, which I and a few other scientists posed independently about 20 years ago: Instead of going through all the trouble of removing phenylalanine from food, why not feed a regular diet that is supplemented with additional amounts of those amino acids (such as tyrosine, tryptophan, methionine, valine, leucine and isoleucine, collectively known as the large neutral amino acids and abbreviated LNAA) whose entry into the brain is impeded by phenylalanine? The higher amounts of these amino acids would increase their ability to compete with phenylalanine for entry into the brain and thereby normalize brain metabolism and development. At best, this amino acid supplement might be able to completely replace the low-

137

phenylalanine dietary treatment for PKU. If it fell short of that goal, it would have a good chance of allowing the treatment to be more relaxed . PKU kids, for instance, would be able to enjoy the luxury of an occasional hamburger or milkshake or pizza, which would likely result in much better compliance than that seen with a very strict diet.

Very recent results have bolstered the hopes for the success of this type of alternate therapy for PKU. A group of scientists in Germany, using a technique known as proton magnetic resonance spectroscopy demonstrated that the technique could measure levels of phenylalanine in the brains of living subjects. They then went on to show that when PKU patients were given a mixture of the large neutral amino acids, this treatment completely blocked the entry into the brain of a dose of phenylalanine. Moreover, the abnormal EEG pattern typical of PKU patients with uncontrolled hyperphenylalaninemia was improved by the supplemental LNAA. These results substantially strengthen the scientific underpinnings of the proposal that treatment of PKU patients with LNAA could prove to be a viable alternative to the low phenylalanine diet. They also provide solid support for the theory that brain damage in PKU is due to the interference by high phenylalanine of the entry into the brain of the LNAA.

A version of this kind of treatment has actually been tested on a small group of older PKU patients (6-20 years) who had, in most cases, discontinued the phenylalanine-restricted diet. The results were encouraging; the treatment improved the neuropsychological test scores, an effect that was presumably related to the finding that feeding the supplemental amino acids also decreased the previously elevated CSF levels of phenylalanine. Given the age of the patients, these signs of success were as much as could have been hoped for. What is needed now are more extensive tests on younger patients to see if the supplemental amino acids are effective in preventing the development of mental retardation. Such studies do not appear to have been carried out. The lack of activity in this area makes it unlikely that this therapeutic alternative to the low-phenylalanine diet will ever be given a clinical trial. As I have previously pointed out, this situation may reflect the inverse of Voltaire's aphorism "Le mieux est l'ennemi du bien": this may be a situation in which "the good is the enemy of the better."

Other ways of altering the low-phenylalanine diet to make it more palatable are being actively explored. One of these approaches is aimed at producing naturally-occurring proteins that have been genetically modified so that they no longer contain any phenylalanine. A protein that has been successfully modified in this way is gamma zein which is found in corn kernels. It contains 203 amino acid residues, two of which are phenylalanine. The gene that codes for the synthesis of this protein has been modified so that other amino acids replace these two phenylalanine residues. In a plant, this modified gene directs the production of phenylalanine-free zein.

This artificial protein after purification from the plant will be used to make a variety of high protein, low phenylalanine foods like breads and cookies. The expectation is that the taste of these foods will be strikingly superior to that of most other low-phenylalanine products. There is a good chance that someday, using methods similar to those used to modify gamma zein, food scientists will be able to make other phenylalanine-free plant proteins. This would open the door to the manufacture of veggie burgers in which both the bun and the burger are made from phenylalanine-free plant proteins.

Work is also under way to produce another potentially very useful phenylalanine-free protein, alpha-lactalbumin (alpha-lac). Whereas this protein is a minor component of cow's milk, it is a major component of human breast milk. Using recombinant DNA technology, the company that brought the world Dolly, the cloned sheep, has already succeeded in replacing the phenylalanine in alpha-lac with other amino acids. Copies of the gene for this modified protein, engineered so that it is only expressed in the mammary gland of a lactating female, have been introduced into the genetic material of newly fertilized cow embryos. When these "transgenic" female cows start to lactate, in addition to normal cow alpha-lac, their milk will contain phenylalanine-free human alpha-lac. It is anticipated that the latter protein will have a much better taste than other phenylalanine-free products and that it will probably be used to relax the strict low-phenylalanine diets that are based on one of the medical foods. It is even conceivable that it can be used to make a delicious phenylalanine-free "milk shake" to go along with a phenylalanine-free veggie burger.

An alternative to the low-phenylalanine diet recently reported by Ania Muntau, Adelbert Roscher and their colleagues in Germany holds promise for the treatment of mild PKU (plasma phenylalanine levels of 600 to 1200 µmole per liter) and mild hyperphenylalaninemia (plasma phenylalanine levels less than 600 µmole per liter). The treatment involves the daily administration of fairly large doses (20 mg per kilogram of body weight) of tetrahydrobiopterin. Treatment with tetrahydrobiopterin was found to significantly decrease the elevated plasma phenylalanine levels and to increase the rate of phenylalanine oxidation in the majority of the patients tested. The ability of tetrahydrobiopterin to decrease plasma phenylalanine levels and to increase the oxidation of the amino acid is, not surprisingly, entirely dependent on the presence of some residual phenylalanine hydroxylase in the patients. As expected, therefore, the treatment did not work on patients with classic PKU, a group that is usually devoid of phenylalanine hydroxylase activity. The treatment also failed to work on some patients with the mild forms of the disease. The failure in this group correlated predominantly with the presence in the hydroxylase of certain mutations known to be located in the catalytic core of the molecule. This latter failure represents a drawback of this new treatment since it inevitably will impose some delay in the initiation of an alternative form of treatment known to be effective, such as the low phenylalanine diet.

Although the cost of treatment of children with tetrahydrobiopterin should not be a major factor, it must be acknowledged that the treatment of older children and adults with the cofactor will prove to be significantly more expensive than the dietary treatment. Based on the price of tetrahydrobiopterin (available from Schircks Laboratories, Jona, Switzerland) in June 2002 of $435/5 grams and a daily dose of 20mg/kilogram, the yearly cost for an adult patient weighing 70 kilograms would be about $44,000. This compares with the estimated cost of the low phenylalanine diet of $5,000 to $10,000 per year.

Having discussed a few of the drawbacks of this new treatment, we must mention one of the most encouraging results. This was the finding that treatment with tetrahydrobiopterin for about 207 days of a group of 5 children with mild PKU, succeeded in increasing their daily phenylalanine tolerance, allowing the patients to discontinue their restricted diets during the test period.

CHAPTER 11

A LOOK AT SELECTED NEWBORN SCREENING PROGRAMS WORLDWIDE

Although, as mentioned earlier, the incidence of PKU varies widely in different countries, the disease knows no boundaries. The problems of early diagnosis and treatment of PKU present a world-wide challenge. Some of the different ways in which this challenge is being met show that progress has been uneven.

The gold standard for PKU screening and treatment is found in Denmark. The tiny size of the country and the comprehensive Danish Health Scheme has allowed the Danes to come up with a unique solution. In 1967, a special treatment home for PKU children- the Kennedy Institute- was established in the town of Glostrup near Copenhagen. The stimulus that ultimately led to the development of this institute was the assassination of President John F. Kennedy in Dallas in November 1963. The Danish people were interested in erecting some kind of permanent monument to the late President and, rather than erecting a statue, they launched a memorial fund aimed at building a combined research institute and treatment center devoted to PKU. The fund received massive support from the press, radio and television. The Kennedy Institute was opened on May 29, 1967, the 50th anniversary of President Kennedy's birth. Over the years, the Institute has expanded its mission to include the study of other diseases that cause mental retardation such as Down syndrome and Fragile X syndrome and has also developed a very strong research program focussed on these diseases

When I visited the Institute some years ago, when it was still focussed on PKU, my guide was Dr. Flemming Güttler, Medical Director of the Institute. During my visit, I was particularly struck by how unlike a hospital or institution it was. It is housed in a one story building surrounding an inner courtyard where children can play. There were many nurses around but they wore what looked like ordinary clothing rather than uniforms. The whole atmosphere was one of wellness rather than illness.

The staff at the Institute consists of a matron and her assistant, both trained nurses, five nursery nurses, two night nurses, two full-time dietitians, three lab technicians, one part-time psychologist, one full-time pediatrician and two full-time medical consultants, one a clinical biochemist and the other a nutritional physiologist. This home serves the entire country.

The Guthrie test is carried out at a single central laboratory on all newborns some time between the 5th and the 7th day after birth . The Kennedy Institute in Glostrup is notified whenever a baby tests positive. The general practitioner is then contacted by telephone and requested to send additional blood samples to both the screening laboratory and to the Kennedy Institute for fluorometric analysis of phenylalanine levels. If this second sample is also positive, with a serum phenylalanine level of at least 0.60mM (about 10 times higher than normal), the family is notified either by their general practitioner or by one of the physicians at the Kennedy Institute. At that point, arrangements are made for the baby to be admitted for a few weeks to the Kennedy Institute, accompanied by the mother, for further evaluation and, if necessary, treatment. The mother can stay in one of the Institute's guest rooms for as long as she can be away from home.

The whole system of dealing with the family of a PKU baby at the Kennedy Institute is structured to cushion them as much as possible from the punishing blow that fate has dealt them. This is a compassionate system designed to minimize the initial shock and sense of despair that parents in other countries have reported experiencing on first learning about their child's diagnosis of PKU. In developing and carrying out the program at the Kennedy Institute, the planners have been guided by John F. Kennedy's philosophy: "Victims of fate should not become victims of our neglect".

During the time when the mother is living at the Institute, attempts are made to build an atmosphere of trust between the Institute staff and the mother. The mothers are informed about PKU and its treatment through frequent conversations and discussions with members of the staff. On a practical level, the mother is taught by a resident dietitian how to deal with the special diet and how to calculate the phenylalanine equivalents of various foods. One of the aims of the interaction between the mother and the dietitian is to teach the mother how to achieve sufficient variety in the low phenylalanine foods so that the dishes do not differ too much from the diet eaten by the other members of the family.

The Institute offers both in-patient and out-patient services 24 hours a day. It also can provide advice to the families and the general practitioner by either phone or letter. This is especially important if the child should become ill, because certain ailments such as infections that are accompanied by fever or vomiting may occasionally necessitate immediate adjustments in the diet. In addition, the Institute can always admit children for varying times, should the dietary treatment prove to be difficult, or if the parents are sick, or on holiday or simply require some relief from the burdens of caring for a PKU child. Many families take advantage of this offer of a break.

As far as how the actual treatment of PKU children at the Institute is handled, when a baby is picked up by the screening program, the mother or, if it is feasible, the whole family comes to the Institute for an initial in-house stay of 2-3 weeks as soon as possible after birth. During this period, the family member(s) are informed about the disease and about how to prepare the diet. The mother is given help in managing the dietary treatment until she feels comfortable handling it by herself. She is also taught how to take blood samples from the baby. Once the child is thriving and the diet preparation and blood sampling have been mastered, the mother and the baby are discharged. At home, samples continue to be taken twice a week during the first year of life, then once a week; later, the frequency is decreased. Samples are sent to the laboratory in the Institute for analysis. Parents are informed about the results of the phenylalanine analyses in writing and are also advised about any adjustments in the diet that might be necessary. Every three months during the first year of life, the child is re-admitted to the Institute for a one or two day stay to monitor progress and for further clinical testing. During the second year of life, the child is seen in the Institute every six months and once a year after that.

Up to the age of 18 years, the patients continue to be treated with the phenylalanine-free formula. After that age they have a choice of staying with the formula or switching to a protein-restricted diet supplemented with amino acid tablets. The tablets, called PreKUnil, contain all the essential amino acids except for phenylalanine and, in particular, are enriched in some of the amino acids that share the same brain amino acid transport carrier system with phenylalanine: tyrosine, tryptophan and the branched-chain amino acids (valine, leucine and isoleucine). The Institute has had 15 years of experience with the use of these tablets and is very satisfied with the results obtained.

As may be recalled, the use of amino acid supplements as an alternate therapy for PKU or as a way of relaxing the strict low phenylalanine dietary treatment was proposed some years ago and has been discussed earlier in this book (see section on Prospects for Alternate Therapy for PKU).

The icing on the cake of this splendid system for dealing with PKU is that every aspect of the diagnosis, care and treatment of the patients is without any cost to the family; it is covered by the Danish Health Scheme. This even includes stays at the Institute, as well as expenses connected with travelling to the Institute and those incurred in providing the special diet once the child goes home.

Since the untreated PKU patient would have to be institutionalized for life (average life expectancy, 67 years), paying for a treatment that keeps PKU patients out of an institution is impressively cost-effective.

Given its great size, it is not surprising that there is no centralized center for the diagnosis and treatment of PKU in the United States. Rather, policies governing newborn screening are determined at the state level. As a result, there is a great deal of variability between states about how the PKU screening program works.

I recently got a glimpse of how the system works in one state, Maryland, when my wife and I visited Mr. and Mrs. K and their two young PKU children at their pleasant home in a small town west of Baltimore. As I expected, both C., a five- year old girl, and her two- year old brother, are beautiful, with light brown hair and light complexions.

The K's learned about "PKU" shortly after the birth of their first child. They learned about it because Maryland law required that before Mrs. K and her newborn baby could leave the hospital their baby had to have a drop of blood drawn by a heel prick to determine whether the baby had PKU. They were therefore prepared when a blood sample actually was drawn from their infant daughter during their 24 hour stay at the hospital. What they were not prepared for, of course, was the shocking news that they learned a few days later, namely, that their baby had tested positive for PKU. They were also told that after discharge from the hospital they should take the baby to their pediatrician's office for a second test for PKU. Because this second test would be done later than the first, it was considered to be more reliable than the first. The second test, however, also came up positive.

They were then advised to take their baby to the Pediatrics Department at Johns Hopkins University School of Medicine where they were seen by Dr. David Valle, a specialist in genetic diseases. During the baby's one week stay at the hospital, they asked Dr. Valle if he would be willing to explain to the whole family (grandparents, aunts and uncles) the nature of the disease, the treatment and the prognosis. The knowledge that they gained from discussions with Dr. Valle helped them cope with the frightening development in their lives.

Partly in response to their personal reactions to learning the devastating news that they have a PKU baby, a group of such families have formed an organization called The Maryland Alliance of PKU Families, Inc. that tries to be helpful to families who have recently learned that their baby has PKU. They send out the following letter to the affected parents:

We are, and we represent, all families in Maryland who are living with PKU. There is no need to sign up and no dues to pay. Through the Maryland State Department of Health and Mental Hygiene, we can contact you and keep you informed of our activities. We are funded through a State grant, and through various activities and donations. It is not our intent to solicit from you. We are here to support you, as much or as little as you would like to receive that support.

You have surely been told by now not to worry about PKU and your baby's future. We know that is difficult to believe when your child has been diagnosed with a disease you have probably never heard of. We have all been where you are now. But we know, and we can assure you from our experiences that your baby will be fine. Listen to your doctors, follow the diet, learn as much as you can, and everything will be fine. We know, because we have been through it and have the beautiful, healthy normal children to prove it.

The Maryland Alliance of PKU Families, Inc. was formed in 1994. By working together, and with our representatives in the state legislature, we were able to have a law passed and signed by Governor Glendenning in 1995 requiring insurance companies to cover the cost of food and formula for the treatment of PKU and other metabolic disorders. It was the first and best law of its kind in the country. This was a significant victory that you can benefit from also.

We work to bring Maryland PKU families together for our mutual benefit. We do this through various activities throughout the year.

We have a party in spring, a picnic in the fall, holiday PKU cookie exchanges, and our main event, a 4-day camp in the summer. We also have quarterly meetings of our board of directors and a quarterly newsletter to keep you and all members informed of our activities.

Our activities allow PKU families to share experiences and learn about the latest food products available. They are a great chance for children to meet and befriend other PKU children.

One other thing is included that we strongly encourage you to use.

That is a list of Maryland PKU families who look forward to receiving a call from you to answer your questions about living with PKU. The families on the list can give you tips you need, tell you a little of what the future holds, and, if you want, introduce you to their children.

Since we do not want to intrude upon this challenging time in your life, we will not call you. If you feel you would like to talk to another family, or find out more about the association, give anyone on the list a call. We are looking forward to hearing from you.

Sincerely,

Members, The Maryland Alliance of

PKU Families, Inc.

There can be no doubt that the activities of this group provide a very useful service to families who have just been informed that they have a PKU baby and are trying to learn how to cope with this devastating news. Instead of feeling that they are strangers in a strange land, they find that they are being welcomed into a new, comforting family.

At some point in the course of our visit with the K family, I asked them whether they were planning to have any more children. They said that they were not. But then Mrs K added that during the time when she was pregnant with their second child, she thought about how she would feel about having a second PKU child. She came to the unexpected conclusion that in some ways she would prefer that the second child would also have phenylketonuria. She felt that the two children would then have a bond and that one would not feel different from his sibling. This statement provided a powerful testimony for how well this family was coping with PKU.

In contrast to the splendid screening and treatment system for PKU in Denmark and the excellent system throughout most of the United States, there are still serious problems in the treatment and management of PKU in developing countries. The experience in Turkey, a country with one of the highest incidences of PKU (385 cases/million births, about three times higher than in the USA), illustrates the magnitude of the problems that cry out to be solved. The information on the system in Turkey was kindly provided by Professor Imran Özalp, Department of Pediatrics in the Faculty of Medicine at Hacettepe University, Ankara, Turkey.

The average age after birth when blood samples are obtained and phenylalanine levels are measured is 15 days. The screening program coverage is approximately 60%. All cases that are positive by the Guthrie test are given a second test in which blood phenylalanine levels are measured by a fluorometric procedure. The time lapse between the initial diagnosis and the initiation of the low phenylalanine diet varies from 7 days to 3 months, with an average of 30 days. The goal of the diet is to keep the blood phenylalanine levels below 6 milligrams/deciliter (360 micromoles/liter) for the first 10 years and around 10-12 milligrams/deciliter thereafter. Blood phenylalanine levels are monitored 2-3 times during the first month, once or twice a month during the first year, and 3 times a year between 2 years and 5 years.

Within the last two years, the screening program has been expanded to include testing for variants caused by defects in the metabolism of tetrahydrobiopterin. Preliminary findings indicate a rather high incidence of this variant form of the disease in Turkey.

The cost of the low-phenylalanine diet is covered for those families who have government or private insurance. Unfortunately, around 40% of the Turkish population does not have any type of health insurance. For those parents who do not have any financial support, the Turkish Foundation for PKU Children provides some help with the supply of the special foods that the children need.

Although blood phenylalanine levels were maintained at a satisfactory level in about half of the PKU children during the first year of life, the percentage of children who were judged to have good metabolic control (i.e., blood phenylalanine levels equal to or less than 300 micromoles/liter) dropped to 27% at the end of the second year. Sadly, only 19% of those who were being treated were found to have good metabolic control throughout the course of the treatment. As a probable consequence of this relatively poor control of the children's blood phenylalanine levels, 37% of the treated children had a mild-to-moderate degree of mental retardation.

It is evident from this survey of the way in which PKU screening and treatment are carried out in Turkey that this country has made great strides in implementing a satisfactory program. But they are not there yet. There is still too much delay in carrying out the Guthrie test as well as in initiating and controlling the dietary treatment. The screening program does not cover a large enough percentage of newborns. And, finally, insurance coverage is inadequate. The results of the screening and treatment program highlight those areas that need improvement.

Whereas there are reasons to be optimistic about improving these deficiencies in the program, there is a pervasive problem in developing countries that will be more difficult to solve: the problem of illiteracy of the mothers of the PKU children. Given the need for long-term compliance with the diet to achieve a satisfactory outcome and the essential role that the parents-especially the mothers- play in carrying out the treatment, it is obvious that parental illiteracy can severely limit the success of the treatment. Also tending to sabotage the program is the high cost of the phenylalanine-free formulas and of the low protein food products for those families that either have no insurance or only limited insurance. Because of these social and economic factors, many of the physicians in Turkey who are trying to cope with these limitations feel that prenatal diagnosis (which is feasible by DNA analysis) and, in contrast to what is done in the USA, the option of abortion of an affected fetus should be integrated into the screening program. There are indications that in Turkey, parents want this option and are not against abortion under these circumstances.

The attitude of this group of parents, which probably also reflects the feelings of parents in other underdeveloped or newly developed counties, is in striking contrast with those of PKU parents in the USA. In the United States, when a physician learns about a pregnancy in such a family, the parents are informed about the availability of prenatal diagnosis, but, because they are not interested in terminating such a pregnancy, with rare exceptions they do not opt for prenatal diagnosis.

The experience in Turkey shows that despite the impressive efforts of the academic community of physicians to apply the latest scientific findings to the detection and treatment of a genetic disease like PKU, the success of their efforts is limited and subverted by two non-scientific factors: the inadequacy of financial support that is available to PKU families and the limited level of education of the parents in these families, most of whom are either illiterate or have only a primary school education.

The experience with PKU screening and treatment in a country like Egypt appears to be very similar to that in Turkey. To put such findings in perspective, it must be emphasized that these countries represent the vanguard in this area of the world. Professor N. Hashem at the Ain-Shams University Medical Center in Cairo, Egypt, a few years ago summarized the situation as follows: "....neonatal screening, therapy and follow-up is either absent or strongly insufficient in several countries of North Africa and Middle Eastern areas."

Even if the ways to remedy the present shortcomings in the PKU screening and treatment programs in this part of the world are obvious- more funds and more awareness of the demands of the treatment- it is difficult to be sanguine about any significant near-term improvement. The reason for this pessimistic view is that in under-developed countries, priority is usually given to other pressing health problems such as malnutrition and infectious diseases. In the fierce competition for funds for the treatment of such conditions that affect large segments of the population, a disease such as PKU, that affects only a tiny fraction of the population, is likely to get short-changed.

CHAPTER 12

HOW DOES HYPERPHENYLALANINEMIA DAMAGE THE DEVELOPING BRAIN?

As already briefly discussed, the first answer to this question proposed that phenylpyruvic acid is the "idiot acid", the compound that interferes with normal brain development and leads to mental retardation. Over the years, this specific proposal has been generalized to include other phenylalanine metabolites like phenylacetic acid and has become known as the "toxic metabolite" theory of brain damage in PKU.

The theory has a lot going for it. It readily explains, for example, why the low phenylalanine diet is an effective treatment: when blood levels of the amino acid are decreased by the diet, the formation of these metabolites also decreases. Moreover, when studied in the test tube, phenylpyruvic acid has been found to inhibit a wide variety of metabolic processes such as glucose oxidation and fat synthesis. This kind of disruption of metabolism could, in theory, interfere with normal brain development. Phenylacetic acid also inhibits certain important transformations in the body.

The devil for the toxic metabolite theory, however, is in the details. The concentrations of these 2 metabolites in blood and CSF of untreated PKU patients are far less-100 fold less-than the amounts needed to inhibit the metabolic processes that have been studied. Indeed, for phenylpyruvic acid, not even a trace amount could be detected in the CSF of 90% of the patients examined. One way to exorcize this particular devil would be to show that the level of these metabolites in brain tissue from PKU patients is 100 times higher than it is in CSF.

The final blow, probably a fatal one, to the toxic metabolite theory came from a very rare variant form of PKU called "chemical PKU" that was reported in two sisters by Wadman and his colleagues. The girls had none of the symptoms of untreated PKU even though they had elevated levels of some of the common metabolites of phenylalanine-phenylpyruvic acid, ortho-hydroxyphenylacetic acid and phenyllactic acid- in their blood. The reason why these girls were protected from the ravishes of the disease are not known.

An alternate and more likely theory than the toxic metabolite theory assumes that phenylalanine, itself, is the toxic substance. In contrast to the "toxic metabolite" theory, this one stands up very well to close scrutiny. Just like phenylpyruvic acid and phenylacetic acid, phenylalanine also inhibits many metabolic processes in the test tube, including protein and lipid synthesis and neurotransmitter synthesis. But unlike the other theory, inhibition is observed at concentrations of phenylalanine that actually occur in tissues from untreated PKU patients. The reason is not that the amino acid is, in general, such a potent inhibitor but rather because the concentrations of phenylalanine are dramatically higher than those of compounds like phenylpyruvic acid.

One of the oldest puzzles in the PKU field concerns the question of why the damage seen in this disease is so specific for the brain. The toxic metabolite theory that assumes that a compound like phenylpyruvic acid is responsible does not provide a good solution to this puzzle. By contrast, the phenylalanine toxicity theory readily solves it. The solution is found in the mechanics of the blood-brain barrier. As may be recalled, this barrier has already appeared to haunt us because it severely restricts the entry of tetrahydrobiopterin into the brain and therefore complicates the treatment of PKU due to a deficiency of the coenzyme.

The barrier also restricts the entry of some of the amino acids, including phenylalanine, tyrosine and tryptophan, into the brain. Problems arise because these amino acids compete with each other for this restricted entry. The problem is compounded because phenylalanine happens to be one of the most aggressive amino acids in gaining entry. It behaves in this situation like the mythical obnoxious character who enters a revolving door behind you but somehow manages to enter the room ahead of you. Because phenylalanine wins this competition, the brain does not get enough of some of the other essential amino acids that it needs for normal protein synthesis, especially during the fetal and early postnatal period when the brain is growing rapidly. As a result, microcephaly is common in untreated PKU patients.

The reason why the brain may be particularly vulnerable to damage mediated by the ability of phenylalanine to interfere with the entry of other amino acids is that the gates guarding the entry of amino acids into other tissues such as skeletal muscle do not appear to be so difficult to bypass as those guarding the brain. If, for example, the same stringent competition occurred in muscle as in brain, untreated PKU patients would likely tend to show signs of muscle atrophy and weakness in addition to being somewhat microcephalic. Finally, powerful support for the notion that brain damage in this disease is due to the ability of phenylalanine to interfere with the uptake of certain other amino acids has been provided by the demonstration that the level of at least one essential amino acid, methionine, is actually much lower than normal in the brains of PKU patients.

Does the gross appearance of the brains of PKU patients look any different from those of healthy individuals? For that matter, is there any correlation at all between the brain's outward appearance and a person's intelligence?

That question has been rattling around ever since Albert Einstein's death. People have been curious about what his brain looked like. Was there anything about the outward appearance of his brain- perhaps the seat of the greatest intellect of all time, (or, as Jeff Schlegel wrote recently in the

Washington Post, "the smartest organ of the 20th century")-to distinguish it from the brains of mere mortals? The answer, disappointingly, is "no". Except for a report by one group of neuroscientists that Einstein's brain was characterized by unusually large inferior parietal lobes, a claim that has been disputed by other brain anatomists, Einstein's brain looked like most other brains. The weight of his brain, for example, was not different from that of controls, a clear indication that a large brain is not a necessary condition for exceptional intellect. This last conclusion is also coherent with data on the brain weights of great writers. It has been noted, for example, that novelist Ivan Turgenev's brain weighed 2012 grams, whereas the brain of author Anatole France was only half that value (1017 grams).

At the other end of the spectrum, scientists had also wondered whether the brain of an untreated PKU patient, one with an unmeasurably low IQ, could, by visual examination, be distinguished from that of a normal brain, or even from Einstein's brain. Other than a moderate reduction in the weight of the PKU brain, the answer, again, appears to be "no".

Whereas casual or even intense visual inspection could not discern any difference between a PKU brain and a normal one, the more penetrating gaze of pathologists and histologists and chemists finally did detect gross chemical and structural abnormalities in PKU brains. In 1950 Alvord and his coworkers reported that most of the PKU patients that they examined at autopsy showed signs of a marked lack of myelin, the substance that serves as an insulating sheath around certain neurons of the nervous system. Based on their findings, these workers suggested that the mental defect in PKU was due to a defect in myelin metabolism.

Alvord's findings, which were amply confirmed by other investigators, highlighted the importance of myelin, a normal constituent of both the central and the peripheral nervous systems. Myelin is a complex substance made up of both proteins (15-30%) and lipids (70-85%). It is the nervous system's insulating material and occurs wrapped around the axons of nerves in the form of what is known as the myelin sheath. It is found mainly in the white matter of the brain, comprising about 50% of the dry weight. By contrast, the gray matter of the brain contains the cell bodies of the nerves and their dendrites, the fibers that connect nerves to each other and conduct the neural impulses away from the cell body of neurons. The insulating properties of myelin are believed to prevent neural impulses from spreading chaotically from one nerve fiber to another, a process that would likely scramble a coherent signal.

Myelin is not a universal component of the nervous system throughout the animal kingdom. Rather, it is a specialized feature of the vertebrate nervous system, where it is believed to facilitate the rapid and energy efficient transmission of large numbers of neural impulses required by these higher organisms. Myelin not only increases the energy efficiency of the transmission of neural signals, but it does it with a great economy of space. The increase in speed of conductance of a neural impulse along an unmyelinated fiber or axon is proportional to the square-root of the axon's diameter, whereas for a myelinated axon it is directly proportional to the diameter. To illustrate what this difference means, for a myelinated axon, an increase in axon diameter from 100 microns to 1000 microns would lead to a 10-fold increase in speed of conductance, whereas for an unmyelinated axon the same increase in diameter would only lead to a 3.3- fold increase. Therefore, to achieve the same speed of transmission with an unmyelinated axon would require about three-times more space than it would for a myelinated axon.

In humans, little or no myelination occurs during the first half of gestation. The process then follows an ancient inborn developmental program: portions of the peripheral nervous system undergo myelination first, then the spinal cord and, finally, the brain. During the first postnatal years, active myelination occurs in most neural tracts with the process continuing during the first decade of life. The last areas of the brain to undergo myelination are those that are involved in higher mental functions such as pattern recognition and reading skills. The process of myelination appears to be integral to function. In general, neural pathways in the nervous system become myelinated before they become completely functional.

There are indications that this relationship is reciprocal, with function stimulating myelination.

The developmental program for myelination is highly relevant to the experience with the dietary treatment for PKU. If defects in myelination do indeed underlie mental deficits in this disease, then it is fortunate that myelination in major areas of the brain occurs postnatally. For instance, the area of the brain in which the neurotransmitter dopamine, known to be deficient in PKU, is synthesized, is the striatum. It is also known that the striatum undergoes myelination entirely postnatally- between the third and twelfth month of life. Had this pattern been otherwise, had striatal neurons undergone myelination entirely prenatally, this vital area of the brain would likely be damaged to some extent by even moderately elevated levels of phenylalanine prior to birth and the low phenylalanine diet probably would be less effective in preventing the loss of dopamine-mediated brain functions.

Studies with one of the animal models for PKU have shown that hyperphenylalaninemia interferes with the synthesis of the protein moiety of myelin, probably by inhibiting the uptake of the other large neutral amino acids. Hyperphenylalaninemia also disrupts the synthesis of the lipid moiety of myelin although it is not clear how it does this.

Defective myelin metabolism in PKU is now firmly established. Most recently, the defect has been confirmed by the sophisticated technique of magnetic resonance imaging

(MRI). Less well established is a causal connection between this defect and the mental retardation seen in PKU. The very same technique that has placed the myelin defect on a very firm footing, MRI, has raised questions about whether this defect has anything to do with the mental retardation that is the hallmark of PKU. The troublesome finding is that the same MRI indications of myelin defects that are seen in untreated PKU patients are also seen in a group of patients who had been treated with the low phenylalanine diet since early infancy and who are not mentally retarded.

The troublesome finding raises the disturbing possibility-disturbing because it would be a setback in our search for the proximate or primary cause of the mental retardation in PKU-that although hyperphenylalaninemia causes mental retardation and also causes defects in myelin metabolism, the myelin defects *are not the cause of the mental retardation.* The conclusion that the myelin defect causes mental retardation would be on the same shaky ground as the one that would be reached if one considered that still another metabolic abnormality caused by hyperphenylalaninemia-dilute pigmentation of the hair and eyes- is the cause of the mental defect. Nobody would seriously propose that blond hair and blue eyes are the *cause* of mental retardation; rather, these are believed to be separate consequences of excess phenylalanine.

Unlike the pigmentation analogy, however, the myelin defect occurs in the nervous system. For this reason, the possibility of a causal connection between defective myelin metabolism and mental retardation remains attractive. But before the connection can be accepted as a causal one, a way must be found to get around the troublesome fact, mentioned above, that myelin defects are still seen in well-treated PKU patients who are not mentally retarded. Otherwise, this may turn out to be an example of a beautiful theory being destroyed by an ugly fact.

In addition to disrupting protein synthesis in the brain, excess phenylalanine also interferes with the formation of critical neurotransmitters. Deprived of its normal supply of tyrosine and tryptophan, the precursors of the neurotransmitters dopamine, noradrenaline and serotonin, the brain does not make enough of these essential molecules. With respect to the synthesis of the neurotransmitters, excess phenylalanine delivers a punishing one-two punch: direct inhibition of the enzymes that make the neurotransmitters and indirect inhibition due to restricted entry into the brain of their precursor amino acids, tyrosine and tryptophan. Like a house with a faulty infrastructure (due to inhibition of protein synthesis) and an inadequate wiring system (due to inhibition of neurotransmitter synthesis), these deficiencies hamper the ability of the brain to perform some of its vital functions and could account for the neuropathology seen in PKU.

A deficiency of dopamine may underlie the reversible deterioration seen in older PKU patients when they go off the phenylalanine-restricted diet or when they switch to a more relaxed diet. The area of the brain most responsible for these negative effects of dopamine depletion has been pinpointed to a structure called the prefrontal cortex. This area of the brain is believed to control "executive functions", a term defined as the ability to maintain an appropriate focus for attaining a future goal. This kind of focus allows for long-term strategic planning, impulse control, and the ability to carry out an organized search, including flexibility of both thought and action.

To measure executive function, subjects are usually given a battery of about 5 different tests. An example of the kind of test that has been used is one that involves a disc-transfer task, known as the "Tower of Hanoi". The test evaluates the ability to plan and carry out a sequence of moves that transforms an initial arrangement of discs to one that is the same as the one arranged by the person giving the test. In preparation for the test, two identical pieces of equipment are set up, one in front of the subject and the other in front of the tester. Each piece of equipment consists of a wooden base with three vertical pegs inserted in it. Three discs differing in color and size are placed on the pegs. The tester's discs are arranged on the right-hand peg to form a tower with the largest disc on the bottom and the smallest disc on the top. The goal for the subject is to arrange his/her discs so they match the disc arrangement on the tester's "tower". Discs have to be moved according to the following three rules: (1) a larger disc cannot be placed on a smaller disc, (2) only one disc can be moved at a time, and (3) the discs have to be on a peg or in the subject's hand at all times. Success on this particular test is very dependent on the subject's ability to do advance planning and the ability to inhibit irrelevant responses (i.e., responses that do not help in arriving at the goal or solution).

When the battery of tests to measure executive function was given to kids with early treated PKU, who were still on the diet, all with IQ scores within the normal range, it was found that their performance varied inversely with their blood phenylalanine levels, i.e., performance was poor when phenylalanine levels were high, but still within what is considered the acceptable range. Since all of the PKU kids were still on the diet, their blood phenylalanine levels were only modestly above normal (3 to 5 times above normal). These results indicate that even mild elevations of the amino acid can lead to deficiencies of dopamine in the prefrontal area of the brain, which can, in turn, impair executive functions.

In addition to decreased levels of dopamine in the prefrontal cortex, levels of this neurotransmitter in the dopaminergic neurons in the retina may also be abnormally low in PKU patients. Such a possibility was raised by findings from earlier studies on patients affected by Parkinson disease, the quintessential illness caused by a deficiency of dopamine.

These patients were shown to have defects in their visual contrast sensitivity that could be traced to their underlying deficiencies of dopamine.

This particular kind of dopamine deficiency can easily be assessed by a simple, non-invasive procedure that measures a person's ability to discriminate low contrast patterns. In a particular version of the test, the subject is shown a group of photos displaying a series of lines that differ in both contrast and spacing. The lines in each photo are oriented to the right, left or straight up and down. The subject is then shown a photo of a test pattern of lines and asked to orient it so that its stripes matched the orientation of the stripes in the test pattern. The ability to discern patterns of low contrast proved to be more difficult for PKU patients than for normal controls, a defect believed to be due to low dopamine levels in the patient's dopamine neurons in the retina.

This test serves as a window for viewing the dopamine status in these particular neurons.

Some researchers have speculated about whether the impaired visual contrast sensitivity seen in PKU patients could be having an adverse effect on the ability of these children to do their homework. In this regard, they have suggested that better illumination and the use of higher contrast material might be helpful in compensating for this problem.

The results of these tests raise serious questions about whether the currently "acceptable" levels of blood phenylalanine are not too high. They also sound a warning about possible subtle dangers that may arise when the low-phenylalanine diet is relaxed too drastically resulting in deficiencies of dopamine in specific areas of the brain.

We have previously mentioned the challenging problem of trying to explain how a genetic defect in the liver- a lack of phenylalanine hydroxylase- specifically damages the developing brain and causes mental retardation. One of the earliest theories to explain this aspect of PKU was put forth by Sam Bessman at the University of Southern California. He latched onto a report by Wapnir and coworkers that brain, especially the fetal brain, has phenylalanine hydroxylase. Furthermore, he asserted in a paper published in 1972 that "PKU is a deficiency of a nonessential amino acid, tyrosine" and proposed that it could be treated by feeding PKU patients with extra amounts of tyrosine. This proposal implied that the disease is caused by the lack of phenylalanine hydroxylase in the brain. In addition, he proposed that the gestational role of phenylalanine hydroxylase is to supply tyrosine to the fetal brain.

This theory had a certain appeal. It avoided all of the complexities of having to deal with the question of how metabolic events in the liver can adversely affect the brain. According to Bessman's theory, there is no need to worry about metabolic events in the liver because the critical events that damage the brain in PKU take place in the brain. The lack of phenylalanine hydroxylase in the brain of PKU patients would mean that the developing brain is deprived of an amino acid, tyrosine, that is essential not only for protein synthesis, but also for tyrosine-derived neurotransmitters like dopamine and noradrenaline.

The theory was refreshingly simple and almost certainly wrong. Wrong because it was based on bad "facts". In my laboratory we were unable to replicate the finding that brain has phenylalanine hydroxylase. Finally, an opportunity to test the notion that tyrosine supplementation without dietary restriction of phenylalanine intake could be used to treat PKU presented itself in 1981. It was shown to be completely ineffective. This negative experience with supplemental dietary tyrosine as the only therapy should not be used as an argument against *the use of tyrosine together with restriction of phenylalanine intake as a treatment for PKU.*

Even without any additional experiments, powerful arguments could have been marshalled against Bessman's assertion that PKU is due to a deficiency of tyrosine. If that were true, the low phenylalanine diet would not be an effective treatment for PKU because the immature brain would already have been damaged before birth due to a lack of tyrosine, secondary to a lack of the brain's supposed endogenous supply of phenylalanine hydroxylase.

CHAPTER 13

HOW CAN METABOLIC EVENTS IN THE PERIPHERY AFFECT BRAIN FUNCTION?

We come now to what is perhaps the most daunting part of the mystery that surrounds PKU, one that is far reaching in its significance: How is a metabolic error that occurs only in the liver- the inability to carry out the normal oxidation of phenylalanine to tyrosine-translated into a defect in the brain's ability to think? How do we go from chemistry to thinking? What is the link that connects these two events?

To place these questions in some kind of broader context, it may be helpful to consider other diseases in which communication between the liver and the brain goes awry. The classical example of such a disease is hepatic coma, known more accurately by the medical term, hepatic encephalopathy (Greek, *encephalos,* brain + *pathos,* illness). It can be considered "classical" because it was first described thousands of years ago in ancient Greece by Hippocrates, the father of medicine.

The connection between liver disease and brain pathology has been forcefully and concisely summed up by the neurologist Fred Plum at Cornell Medical Center as follows- "barring death from accident or unpredictable complications, all

173

progressive liver disease ultimately causes severe cerebral disfunction, with disorders of mentation, motor function and consciousness". In other words, a sick liver leads to a sick brain. Such a dismal linkage is not surprising in view of the recognized general role of the liver in maintaining chemical homeostasis of the organism.

Although the problem has been studied intensively, the mechanism(s) by which an unhealthy liver damages the brain is not known. There is evidence that ammonia, which almost always increases in patients with severe liver disease like cirrhosis, is toxic to the brain. (Liver, during the course of its normal metabolism, produces ammonia, most of it derived from the breakdown of certain amino acids. In the healthy liver the formation of ammonia is usually precisely balanced by reactions that consume the ammonia. But in the diseased liver, this balance is often disrupted with dire consequences. The excess ammonia spills out of the liver and causes mischief in organs like the brain).

In addition to elevated levels of ammonia, levels of some of the amino acids, including tyrosine and phenylalanine, are higher in blood and CSF of patients suffering from hepatic insufficiency. It is likely that these increases will alter the normal ratio of amino acids in the brain, but the pathogenic significance, if any, of such a change in the case of this disease is not known.

Hepatic encephalopathy illustrates well just how sensitive the brain is to metabolic disturbances that originate in the liver, how liver disease can cause metabolic mayhem in the brain. Of course, the underlying cause of the problem in hepatic encephalopathy and PKU are different. In the case of the former condition, the general integrity of the liver is disrupted, i.e., the organ that contains the metabolic machinery breaks down, leading to cerebral disaster, probably mediated by ammonia. In PKU, the organ, i.e., the liver, is not damaged. Rather, a specific step in hepatic metabolism, the conversion of phenylalanine to tyrosine, is affected, an event that indirectly wreaks havoc on the brain.

Returning now to the questions posed in the beginning of this section about how a change in liver metabolism can disrupt brain function in PKU, there is persuasive evidence that the neurotransmitter dopamine holds the key to answering these questions. Dopamine's credentials for playing this role are impressive. As recently summarized by the science writer Sandra Ackerman, dopamine shapes not only our physical functioning in the world but also our ability to process new information, to associate ideas effectively and even to maintain a sense of well- being in balance with realistic perceptions. The way dopamine plays this role is by acting directly on the output neurons of the area of the brain called the prefrontal cortex. These neurons appear to be involved in processing, sorting and assembling information about the outside world. These dopamine neuronal pathways, therefore, are probably the ones by which this neurotransmitter (i.e., dopamine) can influence cognitive function.

We have previously noted that elevated levels of phenylalanine in poorly controlled PKU patients are associated with low levels of dopamine in the prefrontal cortex and the retina. Furthermore, these deficits are thought to interfere with the ability of this group of patients to perform tasks that measure "executive function".

In an earlier section, we have also discussed the mechanism(s) by which elevated blood levels of phenylalanine might interfere with the formation of dopamine in the brain. High concentrations of phenylalanine can, by competing with tyrosine, directly inhibit tyrosine hydroxylase, the enzyme that catalyzes the conversion of tyrosine to dopa, the precursor of dopamine. High blood levels of phenylalanine, working at the blood-brain barrier, can also impede the entry of tyrosine (as well as that of the other large neutral amino acids) into the brain and thereby indirectly inhibit the conversion of tyrosine to dopa and dopamine. It is interesting that both of these mechanisms work by decreasing the amount of tyrosine that is available to tyrosine hydroxylase. These two effects of excess phenylalanine on the rate of formation of dopa could lead to the dopamine deficits that are seen in the brains of PKU patients. Since the normal functioning of dopaminergic neurons depends on them having an adequate amount of dopamine, low levels of this neurotransmitter would interfere with this normal function.

The notion that there is a link between hyperphenylalaninemia and lower levels of dopamine in the brain has been dramatically strengthened by results of recent studies with a mouse model for PKU. A group of Italian neuroscientists has shown that levels of dopamine in various areas of the brain, including the prefrontal cortex, is decreased in this animal model. They also found similar decreases in levels of serotonin.

If inhibition of dopamine synthesis (secondary to inhibition of dopa synthesis) and the resulting shortage of dopamine were the only bad effect of excess phenylalanine, giving extra dopa to PKU patients would be expected to be beneficial (giving dopamine, itself, would not work because this compound, unlike dopa, cannot cross the blood-brain barrier). Dopa has been given for a two week period to a small group of adult PKU patients who had never been treated with the low phenylalanine diet to see if this treatment had any beneficial effects. None was detected.

But inhibition of dopamine synthesis is probably not the only bad effect of excess phenylalanine. In addition to its ability to interfere with dopamine synthesis, high levels of phenylalanine could also interfere with the entry into the brain of the large neutral amino acids and would therefore inhibit protein synthesis. Whereas a large effect on protein synthesis would be expected to lead to a decrease in brain size and function, including cognition, the effect of smaller decreases in synthesis on brain size and function are more difficult to evaluate. As far as brain size is concerned, it is prudent to keep in mind the cautionary note struck by Professor John Dobbing of the University of Manchester, who stated that "It cannot, for example, be assumed that a somewhat smaller brain in a somewhat smaller animal is necessarily a bad thing, unless, for example, one holds to the extraordinary doctrine that the somewhat smaller brain of women is the cause of their alleged mental inferiority to men."

With this caveat in mind, Dobbing goes on to suggest that the physical basis of higher mental activity is probably not related to brain size or the number of brain cells, but rather resides in circuitry- the system of dendritic branching and synaptic connectivity, i.e., the number of connections that any neuron makes with other neurons in the brain. Since branching and connectivity in the brain must ultimately depend on protein synthesis in this organ, it is reasonable to expect that the interference by hyperphenylalaninemia with the entry of large neutral amino acids into the brain and the resulting inhibition of protein synthesis will slow down these essential processes and can therefore contribute to the development of mental retardation. Furthermore, a brain with subnormal branching and synaptic connectivity is probably not in any condition to benefit from extra dopamine. On the other hand, PKU patients who have been successfully treated with a low phenylalanine diet may be further helped by the administration of some extra dopa (or its precursor, tyrosine).

The deleterious effect of hyperphenylalaninemia on dopamine synthesis and, consequently, on the action of dopaminergic neurons, provides an example of how metabolic events in the periphery, such as the liver, could profoundly alter key processes in the brain, extending even to the way we think.

SUMMARY

Professor Charles Scriver, one of the leading authorities on PKU, in describing the history of the disease, has stated that "this has been a long journey of discovery."

It is a journey with many milestones. As already noted, it began almost 70 years ago in Norway when Borgny Egeland, the mother.of two severely retarded children, took them to Professor Asbjörn Fölling at the University Medical Center in Oslo to seek his help. This proved to be a providential choice for Mrs Egeland because Dr Fölling's training in both medicine and chemistry was just what was needed in his attempts to understand what was wrong with the two children. In describing their problem, Mrs Egeland emphasized that in addition to their obvious mental retardation, they had a strange musty or mousy odor clinging to them. This symptom was actually the first hint that they might be suffering from a metabolic disease. This hint proved to be correct!

One of the earliest milestones was Fölling's demonstration that his young patients had a genetic disease that interfered with their ability to metabolize the amino acid phenylalanine. With this pathway blocked, phenylalanine accumulated in the blood and tissues and some of it was diverted into other pathways leading to the formation of substances like phenylpyruvic acid (a phenylketone that inspired the name of the disease as phenylketonuria) and phenylacetic acid. Several years later, in another milestone, Jervis showed more precisely that PKU patients were unable to metabolize phenylalanine to another amino acid, tyrosine, because they lacked an enzyme called phenylalanine hydroxylase. If Fölling is the father of PKU, Jervis is without a doubt the midwife.

For a rare disease, PKU has cast a long shadow, in part because of another important milestone. In 1963 Bickel and his colleagues formulated a low phenylalanine diet, which, when started early enough, was shown to be an effective treatment for PKU, largely preventing the onset of mental retardation that is the hallmark of the disease. In 1963, Frank Lyman, MD, in his book entitled Phenylketonuria, stated that "The prevention of the mental retardation associated with PKU represents one of the great advances in medicine".

Once the need for early treatment was realized, a rapid screening method capable of detecting the disease in early infancy was developed by Guthrie. Currently in America, every newborn infant has a drop of blood drawn by heel prick to test for PKU, usually before the baby leaves the hospital. Such is the reach of this mandated screening procedure that PKU, quite literally, touches every family in America, at least briefly.

The success of the low phenylalanine treatment for PKU had a frightful unintended consequence. Whereas untreated women with the disease were usually retarded and therefore rarely got married or had children, treated women did both. During the years when the details of the low phenylalanine treatment were being worked out, PKU patients were being advised that it was safe to go off the special diet after they had been on it for 6 or 7 years and to resume a normal diet. For PKU women, this advice proved to be disastrous. Unfortunately, most of the babies born to these women were found to have been damaged *in utero* by the mother's high levels of phenylalanine. The abnormalities include microcephaly, congenital heart disease, childhood growth failure and cognitive impairment. It has come to be known as Maternal Phenylketonuria. The condition can be prevented if the PKU woman resumes a low phenylalanine diet prior to conception and sticks to it throughout the pregnancy.

Following the work of Fölling and Jervis, Kaufman and other biochemists showed that the phenylalanine hydroxylating system is complex. In addition to the hydroxylase, itself, the system consists of an essential nonprotein coenzyme called tetrahydrobiopterin and several other essential enzymes, including a reductase that keeps the coenzyme in the active tetrahydro form. Of great significance to our understanding of the underlying cause of PKU, the reductase and the coenzyme were also shown to be essential for the synthesis of the neurotransmitters dopamine and serotonin.

The successful dissection of the hydroxylating system into its component parts turned out to be pivotal, several years later, in unravelling the meaning of the shocking reports that a small group of PKU infants did not seem to benefit from the low phenylalanine diet. Indeed, some of these unresponsive patients died in early infancy. This new variant form of PKU was called "lethal PKU" or "malignant PKU". It was quickly shown that the variants comprised a heterogeneous group of genetic diseases, some caused by a lack of the reductase, others by a lack of the coenzyme, tetrahydrobiopterin. These variants were shown to have a neuronal deficiency of dopamine and serotonin. They could be treated, more or less successfully, by the administration of compounds that can be converted into the neurotransmitters dopamine and serotonin, with or without tetrahydrobiopterin.

Despite the effectiveness of the dietary treatment for classical PKU, the price paid by the family of a PKU child is high. It is paid in the coin of anxiety and the constant burden of preparation of the special diet and the heavy responsibility of making sure that their child's blood phenylalanine levels are kept within the recommended levels. Every one touched by PKU looks forward to a less demanding, less disruptive treatment for the disease. It seems likely that within the next decade or two, gene therapy will be perfected to the point where these families can be relieved of their burden.

GLOSSARY

Allele

One of two or more different forms of a gene.

Amino Acid `

These molecules are the building blocks of
proteins, but they also have other functions in
cells. Phenylalanine, e.g., is an amino acid which
is also a precursor of the neurotransmitter,
norepinephrine. Because phenylalanine cannot
be synthesized by humans, it is further classified
as an essential amino acid. By contrast, tyrosine,
which can be synthesized (from phenylalanine), is
classified as a nonessential amino acid.

Amino Acid Configuration

All amino acids that occur in animal proteins have
an "L" configuration. This configuration is
designated by the capital letter L written before
the name of the amino acid, as in L-tyrosine or L-
phenylalanine. Amino acids of the opposite
configuration occur in some plants and
microorganisms. This configuration is designated
by the capital letter D, as in D-tyrosine. All of the

amino acids mentioned in this book are of the L configuration.

Amino Acid Residue

The generic term for the individual amino acids or "units" in a protein.

Autosomal

Located on or transmitted by an autosome (a chromosome other than a sex chromosome).

Biopsy

Either the process of removing a piece of tissue from a living organism or the piece of tissue obtained by this procedure.

Carbon dioxide

An end-product of metabolism, sometimes referred to as a waste product. Most of this gas is picked up by the capillaries and transported to the lungs where it is exhaled.

Carrier

Also, called a heterozygote. These individuals have or "carry" one normal gene and one abnormal or mutated gene at a specific locus on a chromosome.

Catabolism

The phase of metabolism in which substances in the body are broken down to simpler substances that are often the end-products in a metabolic pathway. The catabolism of phenylalanine involves its conversion to carbon dioxide and water.

Cerebrospinal fluid (CSF)

A liquid, comparable to serum, that circulates through the spaces that surround the brain and the spinal cord.

Chromosomes

Structures in cells on which genes are located. In humans there are 46 nuclear chromosomes (23 pairs) in each cell of the body. One member of each nuclear pair comes from the mother's egg and the other from the father's sperm.

Coenzyme

A substance, usually a small non-protein molecule, which, when combined with a specific enzyme, enhances the enzyme's activity. Also called a "cofactor".

Embryo

The unborn child before the end of the third month.

Enzyme

A substance, usually a protein, produced by living cells that speeds up chemical reactions, leading to the conversion of one substance into another one. Phenylalanine hydroxylase is an enzyme.

Fetus

The unborn child after the end of the third month. Before that time it is called the embryo.

Gene

The physical unit of inheritance, made up of a specific sequence of nucleotides, comprising DNA.

Genetic Code

The sequence of nucleotides in a gene. Groups of three nucleotides, triplets or codons, make up the letters of the code. The genetic code specifies the sequence of amino acids in proteins as they are being synthesized.

Genotype

The genetic constitution of an organism (as distinguished from its physical appearance or phenotype).

Germ Cell

A cell set apart from the other cells in the body which, after union with another cell of the opposite sex, can develop into a new individual. An egg or a sperm cell, distinguished from a somatic cell.

Guthrie Test

A blood test that measures the level of phenylalanine in blood. It is one of the most commonly used tests for screening newborns for PKU.

Hemoglobin

A protein in red blood cells that helps transport oxygen and carbon dioxide.

Homeostasis

A tendency toward the maintenance of a relatively stable or constant internal environment.

Hyperphenylalaninemia

A term that strictly means higher than normal phenylalanine levels in blood. In humans, however, elevated phenylalanine levels are found in most body fluids. The term encompasses PKU, including mild forms, as well as those forms caused by defects in the synthesis and recycling of tetrahydrobiopterin.

Inheritance / heredity

The process by which biological and physical traits are passed from parents to their children.

In Vitro (Latin: in glass)

Refers to experiments carried out in a cell-free system or outside of the whole organism.

In Vivo (Latin: in life)

Within the living body.

Magnetic Resonance Imaging (MRI)

A technique in which radio waves and magnetic fields are used to visualize internal organs and other structures, such as myelin, in the body.

Maternal PKU

A medical condition in which the developing fetus of a women with PKU is damaged by the mother's elevated blood levels of phenylalanine. To prevent damaging the fetus, the woman should be on a low phenylalanine diet, ideally, starting before she becomes pregnant and continuing the diet throughout her pregnancy.

Metabolism

The sum of all of the chemical transformations that occur in an organism.

Microcephaly

Abnormal smallness of the head.

Mutation

An inheritable change in a gene.

Myelin

The white, fatty substance that forms an insulating sheath around certain nerve fibers.

Neonate

A newborn child. Specifically, a child less than one month old.

Neutral Amino Acid

An amino acid that does not have a positive or negative charge. Phenylalanine, tyrosine, and tryptophan are neutral amino acids. These are further classified as "large neutral amino acids." In contrast to amino acids like phenylalanine, amino acids like glutamic acid and arginine bear negative and positive charges, respectively.

Phenylalanine

One of the essential amino acids, i.e., one that humans cannot synthesize. It is normally metabolized, initially, to another amino acid, tyrosine (by phenylalanine hydroxylase) and, ultimately, to carbon dioxide and water.

Phenotype

The observable characteristics of an organism produced by the genotype interacting with the environment.

Plasma

The fraction of blood that remains after the red blood cells or corpuscles have been removed by centrifugation.

Serum

> The fraction of blood that remains after plasma is allowed to clot.

Trimester

> A period of three months.

Trophic

> Pertaining to nutrition or growth.

Units of concentration of metabolites in body fluids

> The concentration in blood of metabolites such as phenylalanine are expressed in various units. Some of the different units commonly used in the literature, including this book, are illustrated below for the normal average phenylalanine concentration in blood or plasma:
>
> 1 mg/100ml; 1mg % ; 58 uM; 58 umol/liter.

Vector

> A construct used to carry recombinant DNA into a cell. Phages are commonly used as vectors.

BIBLIOGRAPHY

Ackerman,S, 1992, Discovering the brain, National Academy Press, Washington, DC

Awiszus, D and Unger, I, Coping with PKU: Results of narrative interviews with parents. 1990 Eur J Pediatr, 149 (Suppl1):S45-S51

Bickel, H, Gerrard J, and Hickmans, EM. The influence of phenylalanine intake on the chemistry and behavior of a phenylketonuric child. Acta Paediatr 1954; 43: 64-77

Buist, NRM, et al., A new amino acid mixture permits new approaches to the treatment of phenylketonuria. Acta Paediatr Suppl 1994, 407: 75-7

Fölling, A, Phenylpyruvic acid as a metabolic anomaly in connection with imbecility. Z Physiol Chem 1934: 227:169-76

Guthrie, R and Susi, A. A simple phenylalanine method for detecting phenylketonuria in large populations of newborn infants. Pediatrics 1963; 32: 338-343

193

Güttler, F, Molecular basis for phenotypic diversity of phenylketonuria. 1992; 3: 5-10

Hanley, WB, et al., Undiagnosed maternal phenylketonuria: The need for prenatal selective screening or case finding. Am. J. Obstet Gynecol. 1999; 180: 986-94

Jervis, GA, Studies on phenylpyruvic oligophrenia: The position of the metabolic error. J Biol Chem 1947;169: 651-56

Kaufman, S, 1977, Phenylketonuria: Biochemical mechanisms. In: Agranoff, BW and Aprison, MH (eds), Advances in Neurochemistry. New York: Plenum Press, pp. 1-132.

Kaufman,S. An evaluation of possible neurotoxicity of metabolites of phenylalanine. J. Pediatr 1989; 114: 895-99

Kaufman, S, 1997, Tetrahydrobiopterin: Basic Biochemistry and Role in Human Disease. The Johns Hopkins University Press, Baltimore, Maryland

Kesby, GJ Repeated adverse fetal outcome in pregnancy complicated by uncontrolled maternal phenylketonuria. J Paediatr Child Health 1999; 35: 499-502

Kure, S., et al., Tetrahydrobiopterin-responsive phenylalanine hydroxylase deficiency. J. Pediatr 1999; 135: 375-378

Lenke, RR and Levy, HL, Maternal phenylketonuria and hyperphenylalaninemia. New Engl J Med 1980; 303: 1202-1208

Muntau, AC, et al., Tetrahydrobiopterin as an alternative treatment for mild phenylketonuria. New Engl J Med 2002; 347: 2122-32

Scriver, CR, Kaufman, S, Eisensmith, R.C and Woo, SLC, 1995, The Hyperphenylalaninemias. In: Scriver, CR, Beaudet, AL, Sly, WS, Valle, D, (eds.) The metabolic & molecular bases of inherited disease. New York: McGraw-Hill, pp 1015-1075

Smith, I and Brenton, DB, 1995 Hyperphenylalaninaemias. In: Fernandes, J, Saudubray, J-M, and Van den Berge, G (eds.), Inborn Metabolic Diseases: Diagnosis and Treatment. Berlin: Springer-Verlag, pp. 147-60

Weglage, J, et al. Psychological and social findings in adolescents with phenylketonuria. Eur J Pediatr 1992 ; 151: 522-25

INDEX

Printed in the United Kingdom
by Lightning Source UK Ltd.
105062UKS00001B

9 781418 437466